Praise for
Is There an Engineer Inside You?
by Celeste Baine

"This book is an excellent resource for a high school career counselor or any student interested in becoming an engineer."

—**The Science Teacher**

"A practical manual exploring the realities of work and career potentials."

—**Midwest Book Review**

"The book provides insights on how to think about an engineering education, get mentally conditioned, and succeed above your classmates."

—**Today's Librarian**

"Useful resource for would-be engineers and engineering students."

—**The Advising Quarterly**

"Book helps students find the engineer within."

—**Engineering Times**

"Students exploring the possibility of an engineering career will find guidance."

—**American Society for Engineering Education, Prism Magazine**

"Features no-nonsense smarts about studying engineering."

—**The NextSTEP Magazine**

"This is a 'must have' book for anyone considering an engineering career."

—**Satisfied Customer**

"Baine provides a realistic look at the skills and training necessary to succeed in engineering and at the great variety of jobs within the field."

—**Parent Press Magazine**

"The book features why getting an engineering degree may be the best thing a person could do for herself."

—**Tech Directions: Linking Education to Careers**

Engineers Make A Difference

Motivating Students to Pursue an Engineering Education

CELESTE BAINE

Engineering Education Service Center

Springfield, OR

Engineers Make a Difference

Motivating Students to Pursue an Engineering Education

by Celeste Baine

Published by:

Engineering Education Service Center (an imprint of Bonamy Publishing)
1004 5th St
Springfield, OR 97477 U.S.A.
(541) 988-1005
www.engineeringedu.com

Printed in the United States of America

Publishers Cataloging-in-Publication Data

Baine, Celeste

Engineers Make a Difference: motivating students to pursue an engineering education / by Celeste Baine.

p. cm.

Includes bibliographical references and index.

ISBN 13: 978-0-9819300-0-8 (pbk.)

1. Engineering—United States I. Title. II. Baine, Celeste.

2. Engineering—Vocational guidance

3. Engineering—Public Opinion—United States

TA160.4C53 2008 620.00973-dc22

How to Order:

Single copies may be ordered from the Engineering Education Service Center, 1004 5th Street, Springfield, OR 97477; telephone (541) 988-1005; Web site: www.engineeringedu.com. Quantity discounts are also available.

Disclaimer

Although the author and publisher have attempted to research all sources exhaustively to ensure the accuracy and completeness of information on the subject matter, the author and publisher assume no responsibility for errors, inaccuracies, omissions, or any other inconsistencies herein.

The purpose of this book is to complement and supplement other texts. You are urged to read all the available literature, learn as much as you can about the field of engineering, and adapt the information to your particular needs.

If you do not wish to be bound by the above statements, you may return this book to the publisher for a full refund.

Acknowledgments

The real thanks for this book goes to my remarkable and generous peer reviewers, Tricia Berry, the Director of the Women in Engineering Program at the University of Texas at Austin and president of WEPAN (Women in Engineering Proactive Network); Tom McGlew, the Instructional Programs Development Specialist for the Maricopa Advanced Technology Center; Cary Sneider, Associate Research Professor, Portland State University, Portland, Oregon; Jo Oshiro, Program Coordinator for the Oregon Pre-Engineering and Applied Sciences Office of the Oregon University System; and Karen Peterson, Principal Investigator for the National Girls Collaborative Project. Without their support and gentle persuasions, this book would not be any more than my ramblings. I am so lucky to have friends of such brilliance.

A big round of applause also goes to my editor, Kristi Koons, for her quick response to my "rush job", and to Amy Siddon for her never ending support, pep talks and patience — especially with my idiosyncrasies.

However, my standing ovation goes to Jo Oshiro for taking my manuscript with her on vacation (I can't thank you enough!) and to Cary Sneider for cheerfully consenting to submit the foreword. Dinner is on me.

And finally, a big thanks goes to every college, school, and organization that has ever invited me to talk to your students or faculty while also sharing your program. Without the conversations and opportunities to learn from you, this book would not be possible.

Credits:
Cover Design: Amy Siddon

This book is dedicated to everyone that works tirelessly to see the spark in a child's eye.

Contents

Appendix

Foreword
By Cary Sneider

Thanks to pioneering educators like Celeste Baine, engineering education is beginning to find a home in classrooms and clubhouses from Maine to California. Educators nationwide are discovering that engineering is just as engaging, fun, educational, and important for students to learn—as science. In fact, it's surprising that it's not the other way around; with engineering part of the educational bedrock, and science looking for a toehold. After all, there are many more jobs for engineers in our country than for scientists, and the engineering processes of invention and innovation underpins our economy. So why does it take a pioneer to bring the obvious to light?

The answer to that question can be found in the delightful chapters of this volume, starting with a colorful introduction by the author about the joyful life of an engineering educator. The book is not a carefully reasoned treatise filled with facts and figures (although there is a good sprinkling of those in the right places), but rather an invitation to explore the many great opportunities to share the fun of engineering with boys and girls of all ages.

Chapter One—The New Face of Engineering—tackles one of the misconceptions about engineering head-on, that it's an all-male geeky occupation. The dozens of interesting engineering occupations, the important role that engineers play in society, and the wide variety people who become engineers quickly dispel the myth, and replace it with a kaleidoscope of intriguing opportunities.

Chapter Two—Girls In Engineering—immediately wows the reader with amazing accomplishments of women engineers, including a famous movie star. The chapter easily transitions to ideas for getting girls interested in engineering.

Chapter Three—Teaching Engineering—starts out with the "whys" and quickly progresses to the "how to's" at different age levels, inside and outside of traditional schools.

Chapter Four—Successful Strategies—offers lots of practical suggestions for how to overcome the stereotypes, build on students' natural interests, and get students engaged in a wide variety of engineering activities in many different settings.

Chapter Five—Successful Curriculum, Programs and Projects—connects with the many other efforts nationwide, to strengthen the foundations of engineering education with well-developed instructional materials so that educators who want to teach engineering don't have to reengineer the wheel.

Chapter Six—Exciting Hooks that Engage Student Interest—provides over a dozen ways that are "guaranteed" to lure the least-interested student to hunger for more engineering design activities.

Chapter Seven—For the Parents of Budding Engineers—encourages parents to get involved and learn as much as they can about engineering. Using the "Ten Strategies to Nurture an Engineer", every parent is sure to find ways to help their child.

Chapter Eight—Recommendations and Conclusions—leaves the reader with new enthusiasm for sharing the fun of engineering education with at least one, and maybe hundreds of tomorrow's budding engineers. Heaven knows we'll need their help in coping with a world population that may exceed ten billion people.

As for the question in the opening paragraph—Why does it take a pioneer to bring the obvious to light? After all, by now there have been enough government reports and gloomy forecasts that if we don't educate more engineers we'll be in big trouble as a nation. So we know where we have to go. But it takes a pioneer like Celeste Baine to make us want to get there.

Cary Sneider
Associate Research Professor
Portland Stand University, Portland, Oregon

Introduction
Purple Cable Ties

One day I was walking through Fry's Electronics, my favorite toy store, to purchase a few computer parts, a wireless network card, toner, CDs and anything else I thought I needed. To my delight, the store was full of things I thought I needed. I spent 2 hours wandering the aisles and filling my arms with merchandise. I purposely didn't get a shopping cart, hoping to curb my spending. As I was convincing myself to make my way to the cashier, I came upon an end cap full of cable ties: black, white, long, short and everything in-between. Finally, my gaze rested on a bag of purple cable ties. Holy moly! I thought to myself, I have to have these ties! This was not a small bag of 10 or 100 or 500 cable ties. This was the mother lode—1000 deep purple cable ties. Right next to it were plain ties and black ones but they didn't do it for me. I had no "need" for plain or black. Did I need cable ties at all? I wondered if I could I honestly justify this purchase.

Earlier in the season, I used my last cable tie on a backyard greenhouse project. To finish the project I had to purchase a little bag of 50 from a local store. They got the job done but they weren't inspiring. They were butter colored. Plain. Blah. Why were these purple ties so important to me? "I'm such a geek" I thought while desperately trying to carry the purple ties along with everything else to the cashier.

I got home and caught myself smiling as I unloaded the bag. These purple ties talked to me and wanted me to use them. They were going to be my friends and every project would be more inspired. I just knew that I would be more successful and that somehow, the manufacturer knew about people like me and made them special; people who need flashing lights, bright colors and "fancy" to capture our imagination and attention.

My Mom tells me that when I was little, I had to have red shoes. It was red shoes or nothing at all. In my red shoes, I believed that I could run faster, jump higher and even do better in school (it helped that before 5 years old, most of your assessments are based on your ability to run around.) The red shoes were a source of expression, prestige and power. Later, in engineering school, I had a special pencil for all of my tests and any difficult assignment. The pencil, just like the red shoes, was a source of expression, prestige and power. By using it, I believed I could score higher....and it worked—most of the time.

Being a person who craves color and works to interest students in engineering careers, purple cable ties became my metaphor for bundling ideas, strategies and inspiration; and also for bringing diversity to engineering education and careers. How many lives could I transform if students could just see the "purple" in engineering?

Without a shadow of a doubt, I believe that "purple cable ties" can help organize our strategies with colorful ideas that will make every student's educational pathway more inspiring and alluring. Especially the segments of the population that are underserved—women, minorities, and students who either don't know what engineering is or who want to be engineers but don't have the role models or the confidence.

This book is about "showing the color" in an engineering career and, as a result, capturing students' passion, imagination, curiosity and dreams; to inspire them to create a life of abundance, meaning and satisfaction from such a pursuit. It's about finding ways to attract diversity in traditionally white, male-dominated fields, and it examines how we can use engineering's full rainbow of choices to enhance the public's perception of engineering—making it more understandable, captivating and socially desirable.

Nightly TV shows such as "Crime Scene Investigation" (CSI) portray medical examiners as intelligent detectives, wearing beautiful clothes, defying dirt stains and having action packed, glamorous work days. Lawyers and doctors also dominate the television airwaves and are portrayed as wise heroes fighting for life or justice. Occasionally, we see genius computer people but they are often the only ones that aren't in the middle of everything. Where are the engineers? Engineering and many technical fields haven't figured out how to make their jobs look sexy and exciting. Hollywood hasn't figured out how to make them appear dynamic, desirable and full of color. But it's coming!

The cable networks are beginning to find large audiences for shows such as "Orange County Choppers", "Myth Busters", "Design Squad", "Smash Lab", "How It's Made", and "Build it Bigger". CBS has a show called "Numbers" that uses math to determine whodunnit. The funding for programs for girls is beginning to pay off with an enrollment surge for women in engineering all across the nation (Guess, 2008). A revolution to prioritize engineering education has earnestly begun and because fewer that 15 percent of students don't see engineering as boring or nerdy (NAE, 2008), the positive message about engineers making a difference will gain footing as all engineering educators continue to build upon the vision of a promising and boundless future. It's just a matter of time and persistence.

As television and cable networks start to catch on to the appeal and significance of engineers' roles in mainstream society, so will the public become as familiar with their fascinating accomplishments in real life. The engineers will be viewed for the pioneering problem-solvers that they are, preventing devastation from hurricanes, exploring other galaxies, and preventing catastrophic illnesses. It will be a world in which people will say, "the engineer saved those people's lives" or "Thank goodness for the engineer!" and the engineers can proudly reflect, "I made a difference!" When these positive messages are conveyed to students, parents, the general public and the media, there will be no stopping what engineers will accomplish in the future.

This book is the result of a decade of visiting engineering schools, talking to engineering faculty and staff, K-12 teachers, counselors, parents and students. As a result of these conversations and presentations I have given nationwide, I have a good perspective on the problems faced in the promotion of engineering as a career choice. Increasing enrollment, retention, attracting the under-represented girls and minorities and motivating students all are addressed within this book. Far from a one-size-fits-all solution, this book supplies multiple recommendations and strategies for you to apply to your particular conditions.

This publication focuses on the state of K-12 engineering education, motivating students, enhancing the public perception of engineering and helping you succeed in developing or sustaining your outreach efforts or program. By using this book, you will expand your knowledge, get ideas on how to thwart stereotypes and rise above the challenges associated with getting students in the door. You will learn to make the connections between everyday things and engineering, by learning what engineers do and how they make a difference in the world around us. Students, infused with your fire and intrigued by the potential opportunities, will see the benefits afforded by an engineering education, and be able to take full advantage of the abundant possibilities for making a difference in the world.

Just as colorful purple ties gave me focus in my work on this book, and in passing along my inspiration to others, so will the color of engineering be apparent in the content of this guide in assisting you and your students as you explore the many career opportunities in the field of engineering.

Some of the ways that I used the purple cable ties to write this book:

1. I bundled all of my ideas about ways to motivate students.
2. I bundled all of my ideas about what it means to be an engineer.
3. I bundled all of my research about the abundant opportunities afforded by an engineering education.
4. I bundled all of the innovative programs about having fun with engineering into a chapter on successful programs.
5. I bundled all of my ideas about why engineering should be taught.
6. I bundled all of the helpful websites I could find into an appendix to help you build your program(s).

Chapter One
The New Face of Engineering

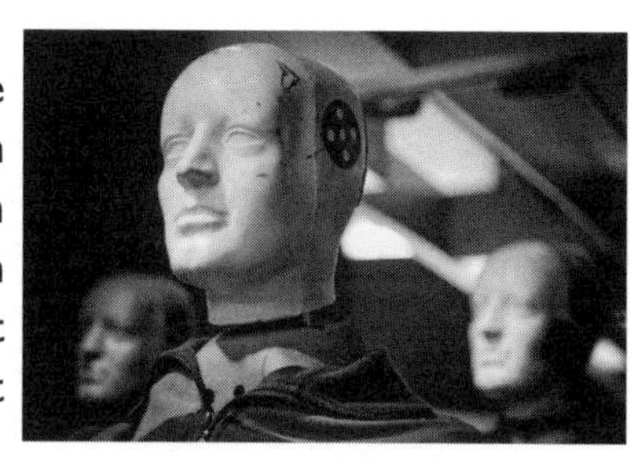

You're a progressive thinker, right? You've let go of the stereotypes about women not doing traditional "male" jobs and math geniuses breaking the fashion rules with pocket protectors. Now, you need to let go of the idea that engineering is all about building bridges and skyscrapers.

Today's engineering majors might find themselves in any of the following scenarios:

- A test engineer crashing expensive sports cars into walls to make them safer
- A forensic engineer evaluating crime scene evidence to narrow down the search for a criminal
- A design engineer creating a robot that can save people from burning buildings
- A pharmaceutical engineer designing or synthesizing a cure for a disease that has killed millions
- A financial engineer analyzing Wall Street's patterns to try to predict future trends

The word "engineer" literally means "one who practices ingenuity." There are droves of people who practice ingenuity – with or without a degree. But the degreed engineers will tell you that engineering is one of the most progressive, challenging, and rewarding fields that can be studied today. Individuals with a bachelor's degree in engineering enjoy some of the highest paychecks of all baccalaureate graduates.

So, what is engineering? According to Jeff Lenard of the American Institute of Chemical Engineers (AIChE), the role of the engineer is perhaps one of the least understood in society. In any poll asking what engineers do, the responses invariably include "fix cars" and "drive trains." We see doctors, lawyers, and police on television, but where are the engineers?

What Do They Do?

Engineers are modern day superheroes and as such, must be ready for anything in an increasingly technology-dependent world. Using math, science, knowledge, and ingenuity in practical ways, they design, invent, create and concoct the most remarkable physical achievements and significant advancements in quality of life known to humanity. They are some of the most creative people on earth. Engineers make the stuff of our lives better, easier, cheaper, more efficient and more fun by solving everyday problems.

Engineers are practical inventors. Through the work of engineers, we are able to have camera phones, wireless computers, satellite TV, airplanes, wind farms, digital music, underwater robots, air conditioning, indoor plumbing, cosmetics, titanium knee and hip replacements. The list goes on and on.

Almost everything you touch has been influenced or designed by an engineer directly or indirectly. It is impossible to think of a major technical development that hasn't included the work of engineers. Many internationally famous companies such as Hewlett Packard, Intel and Apple Computers wouldn't exist if one or more practical inventors (engineers!) hadn't gotten together and made it happen. Solidly rooted in engineering, these companies have grown into giants.

Engineering is a way to make life better. Many problems are solved by applying math principles, but math is just one tool in the engineer's toolbox (We'll cover more about math in Chapter 4, *Successful Strategies*). Inspiration, experimentation, vision, analytical ability, creativity, imagination, energy, passion and communication skills are also extremely important.

If you know a student who wants to reduce pollution, end world hunger, become president of the United States (three presidents were engineers), improve the environment, invent exciting technology, become an astronaut, design race cars, solve complex problems, or be on the cutting edge in a dynamic career, then engineering may be an excellent fit.

Engineers are Creative?

Most people don't describe engineers as creative – in fact, only 3 percent of U.S. adults perceive engineering as a creative career (Harris Interactive, 2004). When you think about engineers being creative, think about it from the standpoint of creative applications. Many people equate the word "creative" with being an artist or writer. Engineers are just like an artist except with a practical twist. They see a problem and apply creativity to find a solution. For example, millions of people all over the world dislike housework. The majority of people would rather be spending time with their friends or family instead of cleaning house. Engineers have addressed and fixed or at least alleviated with a little creativity some of the more time consuming chores.

Engineers are the concept people and often the idea people too. Because consumers decided vacuuming was a problem, now we have the Roomba robot vacuum cleaner that automatically vacuums or mops floors while they do something more enjoyable. New homes often have vacuuming systems already installed in the walls, and Dyson engineers are forever designing a better vacuum cleaner. In fact, a look at the advanced cleaning systems over the last 10 years further indicate just how frequently engineers are employed to find better solutions. Scrubbing Bubbles self-cleans your bathtub daily, the Swifter wants to mop your kitchen, and portable power-washers allow the average consumer to clean the exterior of their home and property without any other special equipment. Without engineers, so many day-to-day chores would be much harder.

On other technology fronts, engineers are the ones who figure out how to make a roller coaster careen forward at 120 M.P.H. in four seconds without killing you. They are the ones who figure out how to make cars that can run on electricity or fuel cell technology to keep

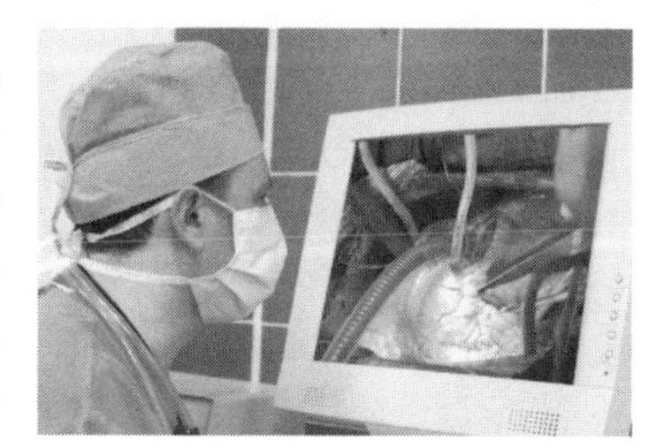

our atmosphere cleaner. They also create medical equipment used by doctors to keep us healthy, and even work in the food industry to make foods such as chocolate and cereal taste better. Engineers have given us music, email and communications that fit into our pocket, and are also hard at work to help save the polar bears and our environment.

In the spirit of developing problem-solving skills, many of the engineering student competitions can also be seen as creative. The American Society of Civil Engineers sponsors a "Concrete Canoe" competition for undergraduate students. The canoe must be made entirely out of concrete; it must float and even race against other concrete canoes created by students at other institutions. In the *Trash to Treasures* competition, students design something useful out of recycled materials and in TOYchallenge, a national toy design challenge for fifth to eighth graders, teams of imaginative kids create a new toy or game. Curiosity, creativity, engagement and dreams are certainly essential elements in each of these competitions.

Problem-solving has been the path by which some of the most amazing inventions and technologies have arrived in the market today. They exist because one engineer had an idea.

Look back at old pictures of the bicycle. People wanted the bicycle to go faster, to go up and down mountains and be more comfortable. The difference now is due to engineering. So that bikes could go off-road and through trails, engineers designed lightweight and stronger frames, along with forks and wheels to take the punishment of off-road riding. When the cost of gas rose so much that more people wanted to ride bikes to work or school, engineers created a lightweight folding bicycle that could be carried into an office and unobtrusively stored away or put into a school locker. When Lance Armstrong needed a faster bike to win the Tour de France, engineers designed that too. Each year, engineers have gone back to the drawing board and made bicycles better. What will bicycles look like in another ten years? It's up to today's students and their imagination to tell us. Today's students will make

the world a better place where people are safer, have more fun and can do more.

What Does an Engineer Look Like?

One of the problems with the public's acceptance of engineering is that it's difficult to say what an engineer looks like. The field of engineering has become larger and more encompassing over time. Biomedical engineering and energy engineering are relatively new. According to the National Academy of Engineering's *Engineer of 2020*, "Because most engineers work in industry and do not interact one-on-one with people who directly benefit from their services, as do physicians and lawyers and teachers, the public is unclear about what most engineers do."

One size does not fit all. So there is not a singular, all encompassing icon that applies. Engineers come in all forms. There are currently 2.2 million engineers, engaged in everything from design to sales to testing, manufacturing, training, and marketing. You can find engineers working in the field, behind a desk, in a production plant, at a customer site, or even on an airplane. Engineers design, manufacture, build, research, write, investigate and present their findings. It's easy to think of engineers designing rides at Disney or crawling around inside of a bridge to check for stress cracks. We know what that looks like but what about the engineers who don't design our modern icons? How do we show an engineer who is checking air quality or researching new and safer ways to dispose of compact fluorescent light bulbs? How do we show the image of an engineer who is trying to find ways to save animals on the brink of extinction? How do we show an engineer who is working on developing safer foods, less hazardous farming techniques or ways to cut down on crime?

The engineering icon also needs to change with the times to be more engaging to students. The men in Mission Control in the movie *Apollo 13* all wore white shirts, crew cuts, and skinny ties. That attire from the 60's wouldn't inspire students today. The IMAX movie *Roving Mars* took a step in the right direction by including colorful and diverse men and women (even if there was only one woman).

Wearing a lab coat, people automatically think doctor or scientist. A headset implies telephone operator, sales person or receptionist. A space suit screams astronaut. Pilots, firefighters, police officers, photographers, teachers, construction workers, farmers and many other occupations have a certain "look". It's a daunting task to come up with just one "look" for an engineer because the field is so broad. As a result, engineers remain less apparent than their inventions and advancements in technology.

Many Roads to Engineering

I live in Springfield, Oregon, the sister city of Eugene and one of the most beautiful areas in the United States. One of my favorite things about this area is that the city has built beautiful, wide, well coordinated bike trails all over the two cities. The system includes 30 miles of off-street paths, 89 miles of on-street bike lanes and five bicycle/ pedestrian bridges. You can get anywhere in Eugene or Springfield by riding a bike.

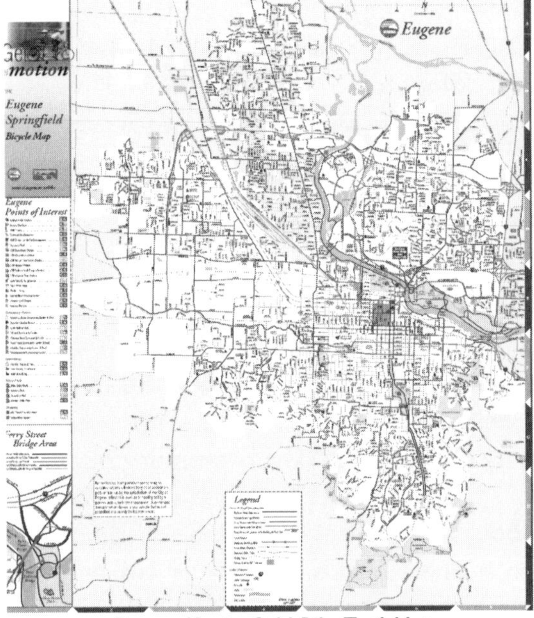

Eugene/Springfield Bike Trail Map

The other day I was called for jury duty and decided to ride my bike. In the five-mile ride, I was one of 32 people on bikes that day (I counted). The trail I took had many places that intersected other trails and many bikers merged on and off at various locations. I equate this experience with engineering education in that there are many paths to becoming an engineer. Not everyone starts at the same place or ends up at the same destination. Not all engineering students are alike; neither is the engineering marketplace. Some students may be able to go very fast and others many enjoy a leisurely pace. Some have racing bikes and others are pulling children in bike strollers. When I'm riding my bike along those trails, I'm never the fastest — many people pass me like I'm sitting still — or the slowest — I pass people too. The important thing for both myself and any pre-engineering student is that we are on the road enjoying the ride, and working toward our destinations.

When I started engineering school, I had never used a power tool or designed anything in a Computer Aided Design (CAD) system. By my second year of engineering school, I felt frustrated because it seemed like all of my male friends could understand the concepts more easily — I thought they had been born with some intuitive understanding about engineering. After a particularly bad Calculus test, I went to my adviser to whine, grovel and basically complain about my inability to "get it". In the midst of my dissertation about all that was wrong with the system, he said, "Celeste, forget about that, the world needs all kinds of engineers." I sat there wondering why I hadn't realized that earlier. I might not ride the fastest — or be a math superstar — but there was a destination out there for me, a destination that fit my personality and one I would enjoy. Engineering is the second largest profession in the nation — with more than 25 major branches and 100 specialties. There is something for everyone who pursues a career in the field. Personal goals, skills, and personality will determine which branch or specialty of engineering is right.

A common misconception about studying engineering is that engineering is only for the intellectual elite or that it's only for students getting A's in math and science in high school. However, there is no standard of intelligence needed to complete a degree in engineering. Anyone who has a genuine interest in taking things apart, solving puzzles and problems, or understanding nature can succeed. As many parents and colleagues can tell you, often students who did not apply themselves in high school find out that when they work on a stimulating problem at an engineering school level, they enjoy the problem-solving process and the solution comes easily. If a student has the aptitude and commitment, he or she can and will succeed in engineering school.

Engineering is Communication and Teamwork

If you have ever played a team sport, you understand that teamwork is integral to the success of the team. Each player brings different strengths to the team, without which the team can't function as efficiently. Engineering design works in the same way. Each member of the team contributes, according to individual strengths, and

the resultant learning and/or design produces a superior product. Jennifer Ocif, a performance footwear engineer at Reebok says, "Communication is a life skill that constantly needs attention and improvement. Unfortunately, it is not specifically taught in engineering classes but you can learn it by doing it anywhere. You just have to work at it because no matter how smart you are, if you can't communicate with the people you work with, your ideas will never go anywhere."

Suppose you have or know of a student that excels at communicating. Traditionally, people would steer that student away from engineering, thinking their skills lie elsewhere. On the contrary, excellent communicators often become the most valued engineers. Those who have excellent verbal, written and people skills, both for technical and nontechnical audiences, who can communicate using technology, and who are intuitive and receptive in understanding the social, cultural and political motivations of people around the world are in great demand. With the solid foundation provided by an engineering degree, an engineering graduate can go anywhere.

Whether an engineering student wants to work at a company employing thousands of people or a smaller company that employs 25 to 30 people, he or she needs to learn to work in teams and communicate effectively. According to Will Pedersen, the Director of Engineering for Rando Productions, "The most creative and best ideas are a result of teamwork, not one man shows." People learn from each other, empower each other, and share the responsibility of finishing the project on time and on budget. Knowledgeable, effective teams can create extraordinary results by tapping into the strengths of each team member.

Engineers must also be able to communicate well with a wide variety of people. Each team member brings a different set of skills to the table. Good communication skills and patience are valuable assets when working with a team of people. It's important to realize that this diversity in ideas and thinking is exactly what can make a product, attraction, or company great. A wide variety of people working together usually equates to more detail and a better design or end result.

Catering to Diverse Cultures

The world population will approach 8 billion people by the year 2020 (CIA, 2006). In the U.S., Hispanic Americans will account for 17 percent of the population, African Americans will account for almost 13 percent and the percentage of White Americans will decrease from 75.6 percent in 2000 to 67.5 percent. By 2050, White Americans will be less than 50 percent of the population (USCB, 2002).

Minorities are important to the engineering profession, and engineering education programs must attract an ethnic and social diversity of students who better reflects the population of the U.S. when they reach their working years. Traditionally, certain sectors of society have not been well represented. Different cultures bring new ideas to the table and the value of such input should be celebrated.

For example, the value of catering to a specific culture can be seen in the popularity of Instant Messaging (IM), text messaging, social networks and chat rooms. These communication styles were popularized by a younger generation that was catering to their differences in thinking and behavior. In AOL alone, 7 billion instant messages are sent each day and 75 percent of online teens are using IM (AP, 2007). The study also found that more teens have used IM for homework than for dating. To meet the needs of a diverse society, we must train a diverse variety of engineers who understand communication preferences and the social and political motivations or movements of cultures around the world.

Ever since a 1970s study by the National Science Foundation showed that ethnic minorities (with the exception of Asian Americans) were vastly underrepresented in engineering, the profession has made efforts to recruit minorities. Various programs now exist to acquaint minority students, their families and teachers with the field as well as to mentor and support minority engineering students. Camps, competitions and scholarships specifically target minorities; colleges have established minorities in engineering or diversity in engineering programs; and many employers have diversity or multicultural departments. Though the rates of enrollment have increased, actual numbers of minority engineers are still low.

Salary

Every now and then when I have the chance to talk to students about engineering careers, I inevitably come across a student who says he or she wants to pursue whatever career makes the most money. At 12, 14 or 16 years old, students aren't sure where their talents lie. They are very bright and only know they want to be as well or better off than their parents. It's ambitious to want more for yourself, and if money can serve as a motivator to obtain good grades in middle and high school, that's a good thing. However, it's also important for these students to realize that money isn't everything and spending every day working in a situation or environment that is not fulfilling will not lead to happiness – no matter how much money they make.

Engineering graduates with a Bachelor of Science (BS) degree earn some of the highest paychecks of all BS graduates. In 2007 many graduates were making $50K to $60K as entry level engineers right out of college. Engineers with a Master's degree (5-7 years of college) earned $65K-$80K right out of college. Although money should never be the sole reason to pursue any career, these statistics are good incentive.

Engineering Attributes

A well-rounded personality is a great attribute to becoming a highly valued and esteemed engineer. To solve the problems of increased population, accelerated global economy, made to order products and environments, health and health care delivery, security, public policy and the public understanding of engineering, the engineer of 2020 will need the following attributes (NAE, 2005):

- Analytical skills
- Practical ingenuity
- Creativity
- Communication & teamwork skills
- Business & management skills
- High ethical standards
- Professionalism

- Leadership, including bridging public policy and technology
- Dynamism/agility/resilience/flexibility
- Lifelong learners

A bachelor's degree in engineering gives a broad knowledge base and leads to a multitude of opportunities. Engineers with a tendency towards right and left brain thinking who are comfortable assessing and taking risks are on the cutting edge in industry, research, consulting, management, teaching, sales, business, and government. Engineering can require a tremendous amount of time and effort, but as technology continues to develop, the need for engineers will increase too.

Chapter Two
Girls in Engineering

Engineers use knowledge, skills and the engineering method to make stuff — tools, structures, processes — to solve problems. They use available resources such as time, materials and labor to do so. As a group, females are more likely to want to use a tool to do something — solve a problem, make a product, streamline a chore — than to want to use the tool for its own sake.

Girls make great engineers! Women have a long history of using tools and materials to solve the problems of feeding, sheltering and clothing their families. For example, weaving can be a highly technical skill, and some of the earliest programmable manufacturing was done on looms.

Some women have gone even further than that:

- ***Ada Byron Lovelace*** collaborated with Charles Babbage, the Englishman credited with inventing the forerunner of the modern computer. She wrote a scientific paper in 1843 that anticipated the development of computer software (including the term software), artificial intelligence, and computer music. The U.S. Department of Defense computer language Ada is named for her.
- ***Amanda Theodosia Jones*** invented the vacuum method of food canning, completely changing the entire food processing industry. Before the 1800's, a woman could not get a patent in her own name. A patent was considered property and women could not own property in most states. So, in a move typical of women inventors of the 19th century, Jones denied the idea came from her inventiveness, but rather from instructions received from her late brother from beyond the grave.
- ***Ellen Swallow Richards*** pioneered the field of environmental engineering with her groundbreaking research into water contamination. In 1870, she helped conduct the first analysis of Massachusetts' water supply and led the research on two subsequent

testings. Her work set the standard for the United States and the world. She showed incredible foresight with her insistence that the earth's environment be examined as a whole, rather than in bits and pieces. She also urged tighter controls over solid waste disposal and air, food, and water purity.

- ***Mary Engle Pennington*** revolutionized food delivery with her invention of an insulated train car cooled with ice beds, allowing the long-distance transportation of perishable food for the first time.

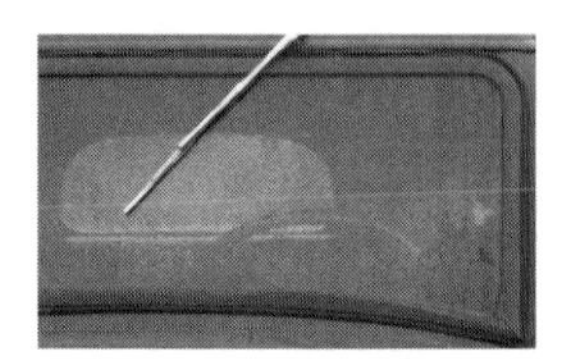

- ***Mary Anderson*** invented the windshield wiper in 1903. By 1916 they were standard equipment on all American cars.

- ***Beulah Louise Henry*** was known as "the Lady Edison" for the many inventions she patented in the 1920's and 1930's. Her inventions included a bobbinless lockstitch sewing machine, a doll with bendable arms, a vacuum ice cream freezer, a doll with a radio inside, and a typewriter that made multiple copies without carbon paper. Henry founded manufacturing companies to produce her creations and made an enormous fortune in the process.

- ***Hedy Lamarr*** was known for her line "Any girl can be glamorous. All you have to do is stand still and look stupid." The 1940's actress invented a sophisticated and unique anti-jamming device for use against Nazi radar. While the U.S. War Department rejected her design, years after her patent had expired, Sylvania adapted the design for a device that today speeds satellite communications around the world. Lamarr received no money, recognition, or credit.

- ***Grace Murray Hopper***, a Rear Admiral in the U.S. Navy, developed the first computer compiler in 1952 and originated the concept that computer programs could be written in English. She once remarked, "No one thought of that earlier because they weren't as lazy as I was." Hopper is also the person who, upon discovering a moth that had jammed the works of an early computer, popularized the term "bug." In 1991, Hopper became the first woman, as an individual, to receive the National Medal of Technology. One of the Navy's destroyers, the U.S.S. Hopper, is named for her.

- ***Stephanie Kwoleks*** discovered a polyamide solvent in 1966 that led to the production of "Kevlar," the crucial component used in canoe hulls, auto bodies and, perhaps most importantly, bulletproof vests.
- ***Ruth Handler*** was best known as the inventor of the Barbie doll, also created the first prosthesis for mastectomy patients.
- ***Dr. Bonnie J. Dunbar*** helped to develop the ceramic tiles that enable the space shuttle to survive re-entry. In 1985, she had an opportunity to test those tiles first hand as an astronaut aboard the shuttle.
- ***Elsa Garmire*** made tremendous advances in optical devices and quantum electronics, making the commercial use of lasers feasible. Garmire discovered and explained key features of light scattering and self-focusing, and a host of other phenomena crucial to optical technology.

Effective Strategies for Engaging Girls

The strategies that are the most effective in engaging girls in science and math (NSF, 2003) are:

- Hands-on activities
- Introduction of role models
- Activities involving teamwork and collaboration
- Activities that have an application to the real world
- Activities that encourage problem-solving

The report also shows that these strategies are not only good for girls — they are good for boys as well. Other career motivators for girls include making a difference in society; having a career that is flexible, enjoyable and rewarding; and knowing that their profession is for "people like me".

If you are trying to motivate girls, it helps to create activities that are naturally appealing to girls. In "Design Squad", students make gadgets such as pancake flippers, and peanut butter and jelly sandwich makers. Many camps focus on creating robotic, technology or toy-making activities. Girls love to create things that make their daily life more interesting or fun. Meaningful projects can include creating kinetic sculptures, bubble or spin art machines, building a personal Etch-a-Sketch, or they can get crazy with Pico Cricket, a programmable robot that helps students create artistic robots.

Here are just a few examples of what engineers can do that might appeal to young girls:

- Create habitats for zoos to keep animals safe and healthy (chemical, mechanical or biomedical engineering).
- Create new medicines and investigate possible cures for diseases such as cancer (chemical, pharmaceutical or biomedical engineering).
- Create a new machine that would allow a blind person to see (biomedical, computer, optical and electrical engineering).
- Using DNA to solve crimes (biomedical, computer, chemical and genetic engineering).
- Find new ways to protect the rainforest (biological, agricultural, civil, environmental, computer, mechanical and electrical engineering).
- Develop techniques to make our favorite foods taste better and stay fresh longer (chemical or food and manufacturing and industrial engineering).
- Create new exhibits and exciting rides for amusement parks (mechanical, civil, structural and electrical engineering).
- Work in the U.S. or in foreign countries to ensure that all people have a safe and healthy water supply (civil and environmental engineering).
- Develop computer programs that help children learn to read, write or communicate (computer, electrical or software engineering).
- Develop new forms of energy to decrease the U.S. dependence on foreign oil (civil, materials, mechanical, electrical, chemical, sustainable and environmental engineering).

The Percentage of Bachelor's Degrees Awarded to Women in 2007 were:

- Environmental Engineering (44.5%)
- Biomedical Engineering (38.2%)
- Chemical Engineering (36.2%)
- Biological and Agricultural Engineering (32.6%)
- Industrial/Manufacturing Engineering (30.9%)
- Materials/Metallurgical Engineering (27.5%).
- Architectural Engineering (24.5%)
- Civil Engineering (21.0%)
- Electrical Engineering (12.4%)
- Mechanical Engineering (12.1%)

Bachelor's Degrees in Engineering By Gender, 2001-2007

YEAR	FEMALE	MALE
2001	19.9%	80.1%
2002	20.9%	79.1%
2003	20.4%	79.6%
2004	20.3%	79.7%
2005	19.5%	80.5%
2006	19.3%	80.7%
2007	18.1%	81.9%

Source: Adapted from ASEE Databytes, 2008

To be effective in promoting engineering to girls, it's important to understand some of the reasons that women turn away from the field:

1. Almost 60 percent of women don't know what engineers do (Harris Interactive, 2004). If a child grows up with a parent, aunt, uncle, friend, neighbor or other relative who is an engineer, they may have a good idea what engineers do. However a large segment of society does not have those role models and may need further guidance.

2. Television often portrays engineering as a male field. Shows such as "Dexter's Laboratory", "Bill Nye the Science Guy" and "Beakman's World" all introduce engineering to girls as if males are intellectually superior. In "Dexter's Laboratory", Dexter is often interrupted by his

> not so bright sister. "Bill Nye the Science Guy" and "Beakman's World" feature outdated, stereotypical men (pocket-protectors, glasses, anti-social, nerdy) as the main characters. The women in the show are usually lab assistants or people who are learning from or looking up to the main character. But things are changing. Two shows on PBS — "DragonFly TV" and "Design Squad" — both have a male and female host. Each episode features teens engaging in different science and engineering activities.

The Wall Street Journal (7/25/2008, A2, Winstein) reported more good news — that girls and boys have roughly the same average scores on state math tests. In the 1970s and 1980s, studies regularly found that high school boys tended to outperform girls. Now, however, in the federally funded study by University of California (UC)-Berkeley and University of Wisconsin (UW)-Madison, researchers found that girls were not outperformed by boys (7/25/2008, A16, Lewin, the New York Times). The findings are based on math scores from seven million students in 10 states, tested in accordance with the Federal No Child Left Behind Act.

If girls are not being outperformed by boys in math, then programs and organizations must focus on opening the doors equally so girls can see all the amazing opportunities in choosing an engineering career. By using teamwork, collaborative problem-solving, and providing hands-on activities that are naturally appealing, they will amaze us with their creativity and independent thinking.

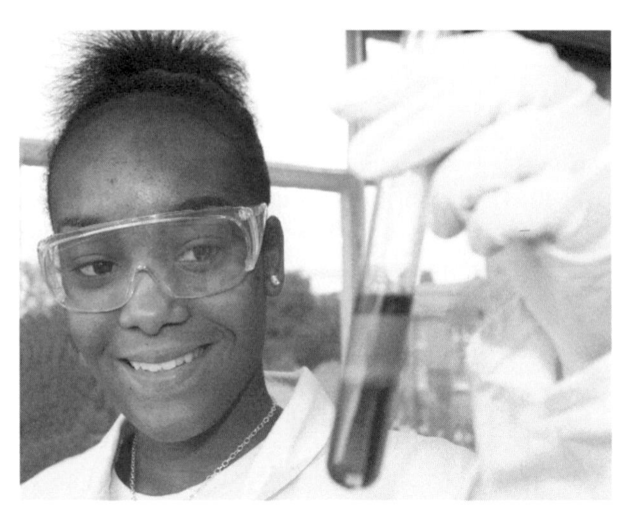

Chapter Three
Teaching Engineering

Engineering is problem-solving. Many teachers enjoy teaching engineering because it combines math and science lessons, team building and creativity with a practical twist. Students learn to work together, increase their communication skills and enhance their presentation abilities by demonstrating and discussing design strategies with the rest of the class.

Students enjoy using the skills and knowledge they have gained abstractly. Engineering projects offer a great venue for students to show themselves and others that they can:

- Manage time and projects
- Study a situation or problem critically
- Research relevant information
- Problem solve
- Design solutions
- Talk intelligently about what they've done and how they did it

Hands-on activities and project-based learning are fun and effective ways to help students learn and retain more math and science concepts. By choosing to teach engineering, teachers can help students make the links between classroom learning, their everyday lives and the wider world. Project based learning can help students visualize abstract science and math concepts. Using hands-on activities, engineering design serves as the bridge to bring color to math and science concepts. This bridge makes our designed world more understandable, relevant and fun. By promoting engineering as a viable career option, teachers also:

1. Help provide a stronger workforce in all fields of Science, Technology, Engineering and Math (STEM).
2. Help create a technologically literate people/society.
3. Provide students with the skills they will need to thrive in a technological society.

How engineering is presented to students can make the difference between them being interested or not. In the old days, when educators described engineers, they called them builders, operators, planners and maintainers. Students today are more likely to respond well to a description of engineers as designers, creators or inventors (NAE, 2008). More exciting messages include:

- Aerospace engineers explore the galaxy!
- Biomedical engineers help people live longer and more comfortably!
- Environmental engineers protect the planet!
- Agricultural engineers feed the world!
- Telecommunications engineers connect the world!

50 Reasons to Teach Engineering

This list could also be called, "50 Reasons to Become an Engineer." They work hand in hand. With a little creativity, any one of these reasons can become a lesson or discussion about engineering careers and serve as a catapult to further exploration.

1. 48 countries (2.8 billion people) could face fresh water shortages by 2025.
2. To save the rainforests.
3. Population in developed countries will age and engineers can help develop assistive technologies so aging people can maintain healthy, productive lifestyles.
4. To give the underserved a clear path to family wage careers.
5. To give students whose talents lie with the concrete rather than the abstract an avenue to success.
6. To make sure students who excel at abstract academics can make the transition to concrete applications and specific problem-solving.
7. To give women another venue for success.
8. To enlighten students who don't know what engineering is about.
9. To save rare or exotic animals from extinction.
10. To educate a potential President of the United States.
11. To help the energy crisis by finding new ways to produce or store solar, wind, wave, geothermal and other sources of energy.
12. To find ways to make nuclear waste non-toxic.
13. To develop safe nuclear energy.
14. To help find a cure for AIDS.
15. To help develop new medicines for numerous diseases.

16. To invent smaller, more affordable computers.
17. To make better theme parks and safer roller coasters.
18. To keep up with the technology needs of society.
19. So the U.S. won't lose all its power to other countries.
20. To give students the tools they need for their futures.
21. To reverse engineer the brain.
22. To counter the violence of terrorists.
23. To improve methods of instruction and learning.
24. To create better virtual reality systems.
25. To capture carbon dioxide.
26. To sustain the infrastructure of cities and living spaces.
27. To explore other galaxies.
28. To understand more about our planet.
29. To reduce our vulnerability to assaults in cyberspace.
30. To prevent devastation from hurricanes and other natural disasters.
31. To improve transportation on land, sea and air.
32. To improve our connectivity and ability to communicate with family and friends.
33. To help us save money on everything.
34. To keep us safe at home and in other countries.
35. To lessen our vulnerability to disease.
36. To improve the quality of the air we breathe.
37. To help our pets live longer.
38. To aid veterinarians in caring for animals.
39. To make food taste better.
40. To make food better for our health.
41. To prevent car accidents with better traffic infrastructure.
42. To create greener buildings and systems that minimize our footprint on the Earth.
43. To understand the oceans and their ability to help us.
44. To reduce the impact of war.
45. To lessen the need for war.
46. To enhance the beauty of our surroundings.
47. To have better furniture and computer peripherals that reduce our risk of carpal tunnel or back pain.
48. To save the polar bears and other endangered species.
49. To get more people where they need to go quickly, safely and conveniently.
50. To decrease the incidence of disease and famine.

The list above can also be used to assign students the task of preparing a report about how engineers are involved in creating a solution for each.

Information Overload

There are hundreds of organizations and colleges that are producing materials to educate students about engineering careers. There is so much information it's difficult to shake out which are the most valuable in serving your educational problem or need. Sometimes I surf the web and get overwhelmed with all the websites, portals and videos on YouTube that tout engineering.

Back when I was a kid, we lived in a time of content scarcity. If we wanted to research something, we went to the library. If we wanted to watch a cartoon, we got up early on Saturday morning. If we wanted to listen to a new song, we waited for the radio to play it again.

Today, kids live in a world of content infinity. When they have a question, they ask Google, Ask.com or Wikipedia. When they want to watch a specific cartoon, they push the "play" button on their on-demand system or they visit the channel's website to watch it on the Internet. When they hear a song they like, they download it from iTunes. They live in a world of made-to-order instant gratification.

For educators, the problem isn't about finding information on engineering careers, locating hands-on activities, or helping students decide which college to attend. It's more about figuring out:

- What is appealing to students (what drives this generation),
- Getting that tailored information to them (books, DVDs, hands-on projects, posters, websites or whatever), and
- Answering the questions that they haven't even thought to ask yet (Will I like engineering? How hard will I have to work?, Is it worth the hard work?, etc.).

Updating the Dialog

The National Academy of Engineering (NAE) conducted a major study to address the messages we convey to pre-college students about engineering. The findings (NAE, 2008) show that young people want jobs that make a difference. Additional recommendations from the research study are as follows:

- Stop reinforcing the images of 'nerdy and boring'.
- Stop focusing on math and science as the needed inputs, and instead focus on the outputs, career opportunities, and making a difference in the world.
- Use the word 'create' not 'build'.
- Use images of people, not things: especially avoid using gears and mechanical looking things.
- Use the following five words in describing engineering: discovery, design, imagination, innovation, contribution.
- Describe engineers as creative problem solvers, essential to health, happiness and safety.
- Emphasize that engineers shape the future.

The recommendations above just require that you update your terminology when talking about engineering.

The other day, when I made a mistake in a conversation, I said, "Sorry, that was my mistake." That same day, when my teenage son made a mistake, he said, "My bad". Same thing, different terminology.

When we saw a man walking who was wearing lots of jewelry, my kids said, "Look at that Bling!" At first I thought the word was derogatory, then they explained it.

Learning a new form of communication is like learning a new language. It takes patience and practice before it sounds and feels right. The important thing is that you keep trying.

Engineering Education Standards

Many state Departments of Education are starting to include engineering education in elementary through high school. Massachusetts is leading the way by including educational content standards in engineering. The states of New York, Pennsylvania, Vermont and Delaware require in-depth learning about the engineering design process that is also found in the Massachusetts framework. Standards in technology with engineering components have also been developed in Arkansas, Florida, Maryland, New Hampshire, and Texas. In fact, Texas has taken this higher education standard one step further by requiring parental authorization for a student to opt out of the college track. A recent state-by-state analysis (Koehler et. al, 2006) found that nearly all state frameworks call for some technology and engineering education with an emphasis on technology and society issues.

The International Technology Association (ITEA) developed about eight excellent standards for engineering education. The standards can be found in their Standards for Technological Literacy publication. Engineering is covered in Chapter 5 – "Design"; Chapter 6 – "Abilities for a Technological World"; and Chapter 7 – "The Designed World."

In chapter 5 on "Design", the objective is for students to develop an understanding of design which includes knowing about:

1. The Attributes of Design:
   - K-2—Everyone can design; and design is a creative process.
   - Grades 3-5—The design process is a way to plan practical solutions; and requirements of design.
   - Middle School—Design is a creative process that leads to useful products and systems; there is no perfect design; and requirements for a design are made up of criteria and constraints.
   - High School—The design process, design problems are usually not clear, design needs to be refined; and the requirements of design.
2. Engineering Design:
   - K-2—The engineering design process; and expressing design ideas to others.

- Grades 3-5—The engineering design process; design is creative; and building models.
- Middle School—Design involves a set of steps; brainstorming is a group problem-solving process; and modeling, testing, evaluating and modifying transform ideas into practical solutions.
- High School—Design principles, influence of personal characteristics; and prototypes, and factors in engineering design.

3. The Role of Troubleshooting, Research and Development, Invention, and Innovation, and Experimentation in Problem-solving: 
   - K-2—Asking questions and making observations; and all products need to be maintained.
   - Grades 3-5—Troubleshooting is a way to fix a problem; invention and innovation turn ideas into real things; and experimentation can solve technological problems.
   - Middle School—Troubleshooting is problem-solving; invention transforms ideas and imagination; and some technological problems are best solved through experimentation.
   - High School—Research and development, researching technological problems, not all problems are technological or can be solved and multidisciplinary approach.

The standards in chapter 6, "Abilities for a Technological World", include being able to:

1. Apply the Design Process:
   - K-2—Solving problems through design; building something; and investigating how things are made.
   - Grades 3-5—Identify and collect information; presenting and selecting the best solution; and testing, evaluating and improving solutions.

- Middle School—Solve problems in and beyond the classroom; specify criteria and constraints; model, sketch or draw the solution; test and evaluate; and document the solution.
- High School—Identifying a design problem, identifying criteria and constraints, refining the design, evaluating the design, developing a product or systems using quality control and reevaluating the final solution.

2. Use and maintain technological products and systems:
    - K-2—Discover how things work; use tools correctly and safely; recognize and use everyday symbols for K-2.
    - Grades 3-5—Follow instructions to assemble a product; use tools correctly and safely; use computers; and communicate ideas.
    - Middle School—Use information provided by manuals, protocols and people; use tools, materials and machines; use computers and calculators; and operate and maintain systems.
    - High School—Document and communicate processes and procedures; diagnose a malfunctioning system; troubleshoot and maintain systems; operate and maintain systems; and use computers to communicate.
3. Assess the impact of products and systems:
    - K-2—Collect information about everyday products; and determine the qualities of a product.
    - Grades 3-5—compare, contrast and classify collected information; investigate and assess technology; and examine trade-offs.
    - Middle School—Design and use instruments to gather data; use the data collected; identify trends in data; interpret and evaluate the accuracy of information.
    - High School—Collect information and judge it's quality; synthesize data to draw conclusions; employ assessment techniques; and design forecasting techniques.

The standards in chapter 7, "The Designed World", the standards were developed for students to gain an understanding of the design world.
This includes selecting and using:

1. Medical technologies (covers developing an awareness of genetic engineering in middle school); and
2. Agricultural and related biotechnologies (covers engineering design and management of ecosystems in high school).

The engineering education standards presented and discussed are visionary and based on the idea that all students should become technologically literate (ITEA, 2008). Living in the 21st century (a technological world) requires that students appreciate technology because it impacts almost every aspect of life. Ensuring that all classroom activities and methods of teaching engineering are standards based is the first step toward providing the best engineering education for K-12 students.

Engineering and Personality

It's important to help students choose a career that is compatible with their personality. Encourage them to find out as much about themselves as possible by:

1. Taking personality assessments at career guidance centers.
2. Talking to their friends, family, guidance counselors, math and science teachers to learn their perceptions about who they are or what they want.
3. Making a list of things they like or dislike.
4. Writing down a description of their perfect job.
5. Writing down their strengths and weaknesses.
6. Thinking about why a certain job is a good fit for their personality or interests.

Have your students check the want ads to see what employers expect, and push them to contact a local college of engineering to see if it offers tours or programs for high school students. For most students, contacting schools or companies will be out of their comfort zone. You may have better success by arranging field trips to colleges or local engineering facilities. Getting parental support is also a great way to manage the situation. You may find that by talking to parents about company or college tours, they'll know of resources or opportunities that you didn't know were available.

Career placement and counseling centers offer the Myers-Briggs Type Indicator®, a primary assessment tool that may give students some insight into who they are, what conditions they may prefer at work, and how they think about things. The test is designed to provide insight on how their interests match up with the interests of others in a particular occupation. However, it should be noted that the assessment, although great for self-discovery, is just an indicator and cannot predict whether or not the student will succeed in the occupation indicated or be a good engineer.

A new online tool for aligning students' interests with careers in engineering is PathAssess from JETS. PathAssess was created from two validated and tested theories of career development and choice: The Holland Theory of Vocational Interest and Lent's Social Cognitive Career Theory. The tool provides each student with their personal Holland Interest Code and compares them to the unique Holland codes of 55 college-level engineering majors. The results are reported in a customized student profile relating their current interests to the types of engineering careers they may wish to pursue.

The engineering personality can be anything:

- Extrovert or introvert.
- Someone who thrives on change, challenge, consistency, or adversity.
- Engineers can be leaders or may prefer to let someone else lead.
- They will probably be hard workers and lifetime learners.
- They may or may not be good under pressure, and may or may not be effective communicators.

There is no "right" personality for a career in engineering just as there is no "right" type of engineer. If students feel it is a good personality fit and are willing to put forth the effort to excel in math and science, then engineering has abundant opportunities. The engineering profession needs all types of engineers and consists of all types of engineers.

When helping students to consider a career in engineering, let them know that it will be a lifelong learning experience; and everything they do to prepare for it will help them reach their intended goal. The more you help expose students to the world of engineering, the more opportunities they can explore.

If You Work With High School Students

Obviously, academic preparation is essential to exploring engineering as a career. In high school, classes in algebra, trigonometry, biology, physics, calculus, chemistry, computer programming, design, or computer applications can be excellent indicators of a student's aptitude and determination to study engineering. While all of these courses are not required to get into every engineering school, early preparation can mean the difference between spending four vs. six years in college. Some universities also require two to three years in a foreign language for admission. Advance placement or honors courses always help, as do high ACT and SAT scores.

But grades aren't the only thing that matter. Students should get involved in engineering-related extracurricular activities to not only beef up their college application, but also to steer their interests and focus within the field. It's an opportunity to decide if they will enjoy the field. Like medicine, law enforcement or teaching, the academic commitment is years long. Volunteering, interning or acquiring a summer job working in a lab, a pharmacy, a manufacturing facility or an architectural/engineering firm will help students figure out if engineering is a good fit. Encourage students to not be afraid to try a few different things until they find something they like.

Formally studying engineering in elementary, middle or high school is a relatively new idea. In fact, according to the Boston Museum of Science, "The high school curriculum we take for granted today was largely shaped by the Committee of Ten, chaired by Harvard President Charles W. Eliot. More than a century ago, the Committee published a definitive report about what all students should learn (Eliot, 1893)."

"The Committee's report called for high school students to study English and mathematics, modern languages, history and geography, and the sciences — physics, astronomy, chemistry, and natural history, which we now call biology. Except for dropping the requirement that all students should study ancient Latin and Greek, the Committee of Ten's report still describes the high school curriculum of today."

When we view the current educational landscape, the news is good. Engineering education is a movement that is gaining momentum as leaders in government, industry and education work together to get back America's footing as a world leader in innovation and technology. Today's students are a critical component of this momentum, and the biggest reason to teach engineering.

If You Work With Middle School Students

Fundamentally, interest in engineering has to be fostered at elementary or middle school level to get students on the right educational track for high school coursework. Preparation for postsecondary education and good jobs begins well before high school (Achieve, 2005). Students who take challenging courses and meet high standards in middle school are much more likely to enter high school ready to succeed. Algebra is widely recognized as a "gateway" course. Students who successfully complete it by the end of 8th grade are much more likely to take rigorous high school courses that will put them on track to earn a college degree.

Middle school is a fun time to teach engineering, and may have the broadest impact for many years into the future. Your job will be much easier if students perceive engineering as fun. Competitions, events, projects, camps and cultivating a can-do spirit will go a long way toward improving the odds that students will be interested. Motivating students at this age is just a matter of being creative enough to tap into their imaginative spirit.

If You Work With Elementary School Students

In elementary school, most students are already engineers. Elementary students are inquisitive, curious, creative and resourceful – just like their older engineering counterparts! Exposing students to engineering from grades 3-8 has the greatest impact. Given hands-on projects and a supportive learning environment coupled with standards-based curriculum, these students are ground shakers. Elementary students seem to find alternative solutions to just about all design problems — so hold onto your hat if this is the age you are engaging. These students will imagine themselves creating new products, developing new technologies and changing the world.

If You Work With Secondary Students Through Outreach

Engineering fairs or one-day coordinated events are where advocates such as engineers, outreach program coordinators, or college level engineering students talk to elementary, middle or high school students about the potential rewards of an engineering career. This can occur on a college campus, in a secondary school classroom or on a field trip to tour a company. Often, this connection occurs in February during Engineers Week but it may happen at any time of year. This is most effective when coupled with a hands-on engineering project, and question and answer time with the students.

To get involved and share engineering with students, advocates must:

1. Find the teachers who are interested and forge relationships. If you forge a successful relationship, the teacher may be able to work before and after the visit to engage the students in activities that maximize everyone's experience.
2. Contact the teachers early in the year. Link your event to something they are already doing in class. Make it easy for them to have you visit.
3. Motivate and inspire the students by bringing projects and giveaways for show and tell. Be prepared to talk about yourself and show your enthusiasm. Students love giveaways such as t-shirts, pins and candy.
4. Support student success by leaving something behind for the teacher. Encourage the teacher to contact you and offer to come back with new and improved resources. Don't forget to call and check in — it's all about relationships.

According to the Business Education Compact, an organization that helps businesses and schools work together, "An engineer who has participated in the program for more than eight years reflected: 'The best part about presenting to classes is seeing the enthusiasm in the students. It is important to get the kids involved and engaged . . .' Another engineer tells of visiting his son's second-grade class. 'In the beginning, I asked how many wanted to be engineers. Maybe 25% raised their hands. At the end of the class I asked how many wanted to be engineers—they all raised their hands.'"

If You Work With Secondary Students by Counseling

Middle and high school counselors play a key role in answering students' and parents' questions about engineering and other STEM careers. Many students highly value their counselor's advice. The challenge and key to advising is to understand the links between engineering and everything else (you can reference *50 Reasons To Teach Engineering on pages 42-44* to help you get started.

Helpful strategies for counseling students about engineering:

1. Learn all you can about engineering.
2. Obtain a library of supporting products.
3. Understand the links between student interests and the world.
4. Show how engineers make a difference.
5. Find profiles of engineers who are doing interesting things.
6. Advise students to take math and science.
7. Advise students to take design classes (even if it's in the art department).
8. Advise students to take as many computer classes as possible.
9. Advise students to take music classes (music and math use the same part of your brain; a huge population of engineers are musicians).

Professional Development

Once you are up to speed in counseling or teaching engineering students, remember that your journey will be filled with learning opportunities. The landscape will continually change as our understanding of engineering education evolves.

Not only is it good for students when advocates come to visit a classroom, it's also good for educators to intern in tech jobs (i.e. during summer sabbaticals) to better understand industry needs. Companies also benefit because they are essentially supporting their future workforce and getting immediate payoff on short-term projects.

Chapter Four
Successful Strategies

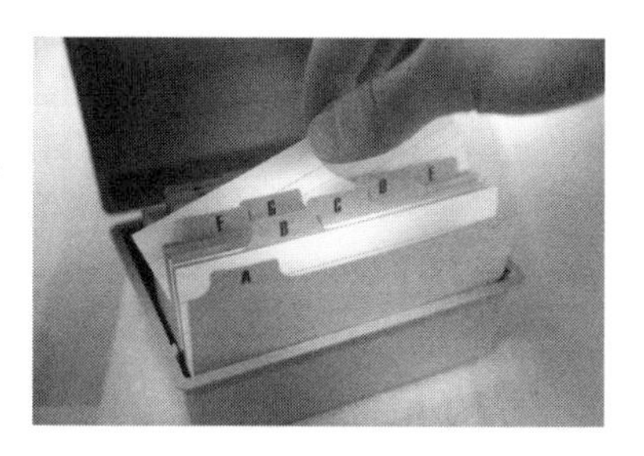

The main objective of this chapter is to help you assist students in finding their motivators. Attracting more students to engineering is critical to successfully increase the number of students who graduate from college with engineering degrees. The strategies listed below are not comprehensive, they are just a compilation of ideas for what will work for many students. Consider each of these a tool to be placed in your toolbox of ideas on how to capture the imagination, dreams, and curiosity of today's youth. A great way to always add to your toolbox is to keep a list, perhaps on index cards, of what works for your students and store them in a recipe box labeled, "Engineering Education Motivators." Whenever you talk to students, receive emails from parents or contact administrators, you'll have an arsenal of information at your fingertips.

Every student has dreams and curiosity that can be captured and used to help them find their way. Your challenge is to find out what makes each of them tick. You need to help them find that little voice inside that says, "YES! This is what I want to do!" When each student finds that inner voice, he or she will be self-motivated and capable of better grades, increased daily motivation, discipline and focus. A career in engineering is a lifelong learning experience, and everything students do to prepare for it will help them reach their intended goal. Every strategy listed below may not work in your various situations but the more ways you find to expose students to the world of engineering, the more opportunities they may have to connect with an engineering career.

Begin by Celebrating

Begin by celebrating engineering classes — to students, to parents and the media. Build an on campus cache of camaraderie and coolness. One of the best events I've attended was a robotics competition. Schools brought their cheerleaders and marching bands to cheer on the dueling robots. It was similar to being at a sports event. It's years later and I can't remember who competed or who won, but I still remember how good it felt to attend.

On the other hand, using the highly competitive sports model may turn-off certain segments of the student population, such as girls, or members of certain cultures. In this case, let the celebration focus on current events in which engineers save lives, help the environment or serve their communities. In Corvallis, Oregon, each July they celebrate art, science and technology with a festival called daVinci Days. It's an amazing festival that helps people understand and celebrate invention and innovation.

When attending the National Science Teachers Association (NSTA) meeting in Boston, I had my picture taken in a space suit. As I stood in front of a white screen, they snapped the picture using a little digital camera and five minutes later, I was an astronaut and ready to go to the moon. The end result was impressive. The picture even had words and logos printed on it so that I could be inspired, and remember the organization that gave me this wonderful memory.

Students love memorabilia. It helps them define who they are by showing where they have been. If you are running an engineering camp, event or program for pre-college students, this can be a form of celebrating. Take a student's picture while he or she is "doing" engineering, put your website or a catchy slogan across it, and the word-of-mouth will spread as that student shows everyone the great photo.

Separate the Boys and Girls When Completing Hands-on Projects.

A few years ago, I was a trainer for the Academy of Engineering, a LEGO laboratory developed by PCS Edventures. I visited schools that purchased the laboratory and showed elementary, middle and high school teachers how to teach math, science and engineering concepts using LEGO. The trainings were marvelous; the teachers were engaged and we all had fun. About halfway through the day, the projects became more open-ended as the teachers were given their instructions on paper, and had to apply creativity to their building. More often than not, this was the point at which one-half to three-fourths of the female teachers "let" their male partner take over, and they became more supervisors than participants. Most of the men were excited to be released from a structured approach to learning whereas the majority of women wanted to take a wait-and-see approach to the new rules. Eventually, at that half-day mark, I split the teachers into teams of males and females and the results were extraordinary. Everyone was building, participating and creating all day long which led to more self-confidence for each participant.

Based on these findings, when first beginning any engineering program, you may see better results by separating the boys and girls. In the early stages of developing spatial analysis skills and learning to build, girls will do much better if you start them out in an all-girl environment. After the girls have been successful and build some self-confidence, team integration will help both genders.

Make Glamorous Associations

From an early age, kids have access to endless content and mental stimulation via cartoons and special effects in movies and on television; and the constant stream of instantly available information, music and current events on the Internet. Within this barrage, kids are shown representations of various professions. They see doctors and lawyers all the time but where are the engineers? The few images they do see of engineers in the media are not clearly or accurately defined. To get the kids of today interested in an engineering career, we need to make the pop connections with images of engineers that resonate on the same level as that of the more familiar images of doctors and lawyers.

As an assignment to get the students thinking, have them watch a movie that you select and identify the engineers doing engineering work. Three examples are listed below. See how many more you can come up with.

a. Show students how Dr. Octopus in *Spiderman* was an engineer instead of a mad scientist. Have students build a list of all the reasons why Dr. Octopus was an engineer (Peter Parker is also an engineer but they never call him one).

b. Willy Wonka was a chemical engineer. Have students watch this movie in class or as a homework assignment, and identify scenes in which he did engineering work; and have them identify what kind of engineering.

c. Identify the simple machines in *Pirates of the Caribbean*. Talk about who made these machines and see if students can find the inventor or first documented use.

Celebrity Engineers

To add to the glamour, here are some stories about famous people you may not have known were engineers or that they had engineering backgrounds. This is a follow up list to chapter two's Girls in Engineering.

- **Thomas Edison** created the first practical incandescent lamp in 1879; it used carbonized cotton thread that glowed for 40 hours. He also invented the phonograph and the world's first central electric power station in the 1870s. To this day, he is credited with more than 1,000 patents.
- **Cindy Crawford**, a famous international model and actress, graduated from high school as valedictorian with a 4.0 GPA. She won an academic scholarship to study chemical engineering at Northwestern University.
- **Jimmy Carter**, the 39th president of the United States, attended Georgia Tech and the U.S. Naval Academy. He served in the navy as an engineer working with nuclear-powered submarines, and later retired to manage the family's peanut farming business.
- **Alexander Graham Bell** invented the telephone. Upon Bell's death, all telephones throughout the United States "stilled their ringing for a silent minute in tribute to the man whose yearning to communicate made them possible."
- **Herbert Hoover** was the 31st president of the United States. He graduated from Stanford University in 1891 with a degree in

mining engineering. He went to China and worked for a private corporation as China's leading engineer.

- **William Hewlett** and **David Packard** met as undergraduates at Stanford University. The two engineering classmates founded Hewlett-Packard in a small garage in 1939. They started their multi-billion dollar company with an investment of $538.
- **Elijah McCoy** invented a lubricator for steam engines. His invention revolutionized the industrial machine industry because it allowed machines to be oiled while still in motion. The term "real McCoy" refers to his invention. It became so popular that people inspecting machines would ask if the machine contained the real McCoy, meaning the real thing. Elijah McCoy was educated in mechanical engineering in Scotland and is credited with other inventions such as the ironing board and the lawn sprinkler.
- **Alfred Hitchcock**, film director of *Psycho* and *The Birds,* received a mechanical engineering degree from the London County Council School of Engineering and Navigation in Poplar, London, England.
- **Montel Williams**, author and syndicated talk show host graduated from the United States Naval Academy with a degree in engineering.
- **Aileen Schumacher**, writer of engineering mystery books, is a civil engineer.
- **Steve Wozniak,** co-founder of Apple Computer, is a computer engineer.
- **Neil Armstrong**, the first person to set foot on the moon, is an aerospace engineer.
- **Dolph Lundgren**, actor in action hero type movies such as *Masters of the Universe, Missionary Man,* and *Command Performance* has a master's degree in chemical engineering.

Find out what students like and link it back to engineering.

This is a great way to connect with students because you can link anything to engineering!

a. Do they like music? If so, have them research and write a paper about the engineers who developed the iPod. Or, have them give a presentation about how CDs, DVDs, digital instruments, headphones or microphones are made.

b. Do they like skateboarding or sports? Bring a skateboard to class and have the students think about the design of the trucks, bearings or board. Show what things a skateboard engineer would do. How would your students do it differently?

c. Do they like rock climbing? Bring the equipment to class and have students think about the various materials that make this sport possible.

d. Do they like clothing? This can lead to great discussions about modern fabric performance. Companies such as Nike have many engineers that work as textile engineers trying to find materials that wick moisture away from the skin, during exercise or materials that keep muscles warm without bulk.

e. Do they like to eat? Show them how engineers work at Hershey, Mars, and other chocolate factories developing delicious chocolate treats.

f. Do they enjoy video games? Howstuffworks.com has videos about how games such as Guitar Hero were created.

g. Do they enjoy watching movies?

h. Do they enjoy using a cell phone?

i. Do they like to chat online?

Information for students to give reports on their favorite products is available at How StuffWorks (howstuffworks.com). If a student is particularly interested in video games, it is always good to have students include an employer's (Nintendo,

Playstation, EA games, etc.) job posting for an engineer. You will find that many postings require excellent communication skills as well as technical ability. For example, to become an iPod engineer requires technical ability, creativity, communication skills, passion, energy and the ability to work in a team (www.apple.com). Most students don't realize that being a well-rounded person is a good skill for engineers!

Whatever students are interested in — medicine, robotics, architecture, design or science — there's probably an engineering job with their name on it. Of course, by the time they graduate from college, there could be a whole new subcategory of engineering. They could be working on robots that can cook gourmet meals, building skyscrapers on Mars, or even providing the technology for performing brain transplants.

Become Well Informed

If you don't know why engineering education is gaining momentum and how engineers make a difference in the world, you need to get informed. This book is a good start but is just one source of information. Subscribe to blogs, read magazines and articles on engineering education and find out what's going on in your community, college of engineering and schools. Become a part of engineering education organizations such as the American Society for Engineering Education (ASEE) or the International Technology Education Association (ITEA).

There are many publications on engineering careers that can help students when they want to make career choices. How to choose an engineering major is a huge decision. See the appendix for a list of web links, publications and helpful organizations.

Provide Mentors

Another excellent approach to helping students become an engineer is to establish a mentor network. Mentoring is successful because it's a one-on-one learning experience that can be so much more than a technical learning experience. Mentors can help students learn approaches into competitive industries, help them network, introduce them to key players, teach them how to listen, and help them evaluate solutions to problems. Mentoring is a part of being successful in any industry but especially for careers that are competitive.

According to the Advocates for Women in Science, Engineering, and Mathematics (AWSEM), "evidence overwhelmingly suggests that girls beginning their exploration of science and math, as well as women who have already achieved high career goals in these fields benefit tremendously from vertical, dynamic mentoring networks." Both groups profit from such mentoring relationships as both the more experienced individual and the less experienced individual gain from each others insights, experiences, and enthusiasm.

There are many ways to find mentors. Encourage students to consider questions such as: "Who can I talk to about my career?" " Who will take a special interest in my goals?" " Who do I admire?" " Who do I want to model myself after?" "Who do I want to emulate?" Have them try to pick a mentor in their field of interest, but don't be limited by this approach. The most important thing is that they find someone they respect, admire, and can talk to easily.

Sometimes, it's hard to imagine a student in elementary or middle school having a mentor but if done right, it's very effective. Baldwinsville Central School District in Baldwinsville, NY has implemented a "Girls in Engineering Day". For this day, they have 40 to 80 middle school female students sign up with their middle school counselor prior to the event. They then have all the female students currently enrolled in the high school engineering courses act as mentors and role models for the entire event. Several engineers come in on that day and run three to four activities for the middle school girls. The activities are typically open-ended design problems that, once solved, can make a difference in peoples lives. Over the last six years, the district has seen its female enrollment double and then triple for the first year engineering class offered in high school.

Network

Networking with other engineering educators is also helpful. Find out about programs at other schools and what other educators are doing to motivate students. This is a great opportunity to learn the nuts and bolts of launching new programs and successfully recruiting students. Build relationships with the dean and/or teaching staff at your local community colleges and college of engineering to find out how to get in the loop. Find out about career days and tour days, and arrange to have engineering students bring their projects when they come to talk to your class.

Encourage students to also network. They should begin to talk to every engineer or engineering student they know about the challenges ahead and how to prepare for them. Students can also network with other pre-engineering students in MySpace or Facebook. They can find friends, join pre-engineering or technology focus groups, find out what other schools are doing and have fun. MySpace and Facebook have groups for FIRST Robotics, JETS, ASCE, IEEE, Engineers Without Borders, and more.

Engineers Week

Engineers' Week, or eweek, is a unique time that takes place every February during the week of President's Day. Reports from Engineers Week 2006 state that 40,000 engineers visited classrooms to promote the profession, and to give students a better understanding of engineering. This is the engineers' chance to show the academic world that engineering is an exciting career, and that engineers really can do anything!

To have an engineer come to your class, event, or program, call local engineering firms, local engineering society branch offices (IEEE, SWE, ASME, etc.) or local colleges or universities to ask about possible activities, competitions or events. Get involved. Another useful eweek activity is to convene a panel of engineers to speak over lunch with students on how they became engineers. This is less formal and can therefore be more engaging and fun for students.

In Oregon, the bulk of engineers are matched to classrooms via the Business Education Compact (BEC). The BEC is a community service organization that provides programs to connect Oregon businesses and industries with schools. This type of organization is not unique to Oregon and may exist within your state as well. The best approach to finding a similar service is to do an Internet search for the local branch of such organizations.

When planning events for Engineers Week, don't overlook the Society of Women Engineers (SWE). If you are a teacher or event planner, send an email to your local branch asking for an engineer to come talk to your class or participants. Not only is this great for all students but it is especially good for girls to see women engineers and understand the profession from a woman's viewpoint. In addition, from the contacts you make, you may be able to arrange for some students to job shadow. Many SWE sections are already doing outreach activities. As a result, you may find out that your local SWE section is hosting events throughout the year.

Summer Camps

Summer camps provide another innovative approach to helping students prepare for a career in engineering, or evaluate if that career is a good fit. Students can find out what it's like to study engineering, the different types of engineers, and what engineers do on a daily basis. Many universities in the United States offer residential and/or commuter summer engineering camps for middle or high school students. They can help students develop leadership, professional and personal organizational skills, and opportunities to meet and talk with engineers during visits to local engineering companies. Check with the college of engineering at a university near you to see if any summer programs are a good fit for your students. Or, visit *www.engineeringedu.com* to help students find a camp in your area.

Student Competitions

A great way for students to get a feel for engineering is to look at the student design competitions that are sponsored or co-sponsored by various engineering societies and organizations. These competitions are developed to encourage and motivate students. The competitions focus on teamwork and allow the students to get a "real world" feeling for the design process, including cost of materials, and experiencing a team dynamics environment.

A few of the more popular competitions include:

- **National Junior Solar Sprint**. A U.S. Department of Energy program where student teams in grades 6-8 construct model solar-powered cars and race them. *www.nrel.gov/education/student/natjss.html*
- **FIRST Robotics Programs**. Includes programs for elementary through high school students in robotics. *www.usfirst.org*
- **Mathcounts.** A national math coaching and competition program for 7th and 8th grade students. *www.mathcounts.org*
- **Future City**. Students learn about math and science in a challenging and interesting way through reality-based education using SimCity 2000 software. *www.futurecity.org*

- **TEAMS and NEDC**. Academic and design competitions that challenge high school students with real world engineering issues. Sponsored by the Junior Engineering Technical Society (JETS). *www.jets.org*
- **Design Squad**. The Design Squad Trash to Treasure Competition challenges kids to take everyday discarded or recycled materials and re-engineer them into functional products. The grand prize winner received a $10,000 cash prize provided by the Intel Foundation and a trip to the development lab at Continuum, an award-winning international design and innovation consultancy, to build a prototype of his or her Trash to Treasure design. The competition is for students who are between the ages of 5 and 19 (and not have graduated high school) at the time an entry is submitted. *www.designsquad.org*
- **TOYChallenge.** TOYchallenge is a national toy design challenge for 5th-8th graders and a chance for teams of imaginative kids to create a new toy or game. Toys are a great way to learn about science, engineering, and the design process. That's why astronaut Sally Ride brought Hasbro, Sigma Xi, Southwest Airlines, and Sally Ride Science together to launch this challenge! *www.toychallenge.com*

Check with the college of engineering at a university near you to see if any competitions are offered for local secondary schools. Or, visit *www.engineeringedu.com* to help students find a competition in your area.

Clubs

Another approach to having your students find out if they will like engineering is to have them join a science or engineering club in your school. This can be a way for them to have fun with math and science, meet other students interested in those subjects, and learn about amazing careers. A club can provide boys or girls with hands-on science, engineering, and math activities that are both fun and educational. Club activities can also include visiting local companies to meet and learn about men and women in science, math, and engineering careers.

How to Talk About Math

On an airplane earlier this year, I exchanged greetings with the guy next to me. Eventually we began talking about what we each do for a living. I told him I was an engineer and that going to engineering school transformed my life for the better. As a result, I now spend my life encouraging kids to explore all the opportunities in engineering. His eyes lit up and he said that he also had wanted to be an engineer but could not get through the math. I asked him what he does today and he told me he is an investment banker. I sat there stunned, staring at him in a very confused state with my mouth hanging open. "Don't they teach math in business school? Don't you have to be friendly with numbers as an investment banker?" I asked. He responded that he had many math classes in college, but he thought business math was easier than engineering math.

As many of you know and probably struggle against, math is one of the gatekeepers. Perceptions about math have changed the course of millions of lives. Sometimes all it takes is a bad teacher in first or second grade to change a student's direction, and sometimes it's about the challenge of many details — like changing a minus sign to plus when putting it on the other side of the equal sign. Sometimes, tending to the details of math (and life) can seem overwhelming.

But, we need to tell students that math is just one tool in the engineer's box. Math and science are important tools to understanding the world and getting through most engineering classes, but they are not the only tools that an engineer uses to solve problems. Fortunately, in engineering there are thousands of different types of jobs. A student can choose a job that is very math intensive or a job that prioritizes and uses different tools in problem-solving.

The important thing is to learn why and when math should be applied, and to know what the approximate answer should be if entering an equation in a calculator or computer. For example, if an engineer enters an equation into a computer, he or she should have an idea what the answer will be so they can know if the equation was entered correctly. If the computer doesn't output an "expected" answer, that may tell the engineer an assumption or part of an equation might not be correct.

However, it's also important to encourage students to take as much math as possible in school. Without at least three years of high school math, students will be excluded from a wide variety of jobs including: Engineer, Programmer, Accountant, Biologist, Medical Technician, Architect and Doctor. Math is very important for intellectual development including creativity, constructive processes and problem-solving (Singh, Man pal, 2005).

In most schools, the key "math decision-making times" are:

- 8th grade, when most students decide if they will take algebra in 9th grade, an important first step to continued math involvement (although more advanced students may take algebra in 8th grade) and
- 11th grade, when math requirements for college admission are fulfilled.

Kyle Milliken, a mechanical engineer for an international manufacturing facility in Huizhoua, China said, "To me, engineering is not a job but a way of seeing things and can lead to literally a world of opportunities. This way of seeing things or perspective is the key ingredient to being an engineer. Walt Disney wants a new thriller ride to delightfully terrify visitors at the park, some person grabs a piece of paper and starts sketching. A million pounds of aluminum, jet fuel, people and tasteless food goes rumbling down the runway and takes flight. It's because some person had an idea. A doctor inserts a tiny device into the beating heart of a patient and his or her life becomes better; it's because somebody had a concept. Engineers are those people. Engineers turn ideas into reality. The vision, the ideas and the concepts are the important part. Math is just a tool that engineers use to do it."

There is something for everyone in engineering. Every student can personally design his or her future. Other tools in the box include creativity, passion, communication skills, teamwork skills, common sense, analytical ability, writing skills, presentation skills and time management.

Use Music to Build an Emotional Connection

One summer I tuned into the Live Earth concerts (promoting "green" awareness to save the planet), and heard a report that discussed the reasons why they use music to build consciousness about global issues. The announcer said they use music because music builds an emotional connection. It is through this emotional connection that people will hear a message that they may not be able to hear otherwise.

Wonderful idea! Do it with engineering too. Watch almost any commercial on TV and you'll hear music in the background trying to elicit some emotional response from the viewer. Kids are bombarded with so many stimuli (noise, color and movement) that by the time they get to class, they are tired and may have trouble concentrating in a quiet environment. Their world rarely lets them be quiet. And now that so many middle and high school kids have cell phones, the odds of having quiet time have seriously diminished even further.

As educators, we can begin the battle by trying to relay information to students in ways they are more accustomed. For example, if you can integrate music into your everyday lessons, you may reach a few more students than you might otherwise.

Here are a few examples:

- Put soundtracks into your PowerPoint presentations.
- If you hold a competition such as Bridge Building, Future City, Robotics, or Rube Goldberg, blast music in the background to build team spirit and an emotional connection between the students and their engineering project.
- If your class is rowdy and unfocused sometimes, playing music can make the students settle down and focus on the lesson or coursework instead of on each other.

Use Graphics to Enhance Meaning

Graphics are a form of communication often overlooked. Communication, especially in engineering, can mean the difference between getting the job or not. In this age of instant and text messaging, students need to take every opportunity to enhance their communication skills.

Engineers Week has an activity posted called, "The Microprocessor: Peanut Butter and Jelly Activity." In this activity, students create a precise set of instructions to make a peanut butter and jelly sandwich. This is an excellent communication activity because after the students work on writing the instructions, you can have them redo the activity using text and illustrations or just with illustrations.

Each fall, The Engineering Education Service Center hosts an engineering poster contest to allow students to visually communicate their ideas about an engineering theme. Past contests with themes such as "Engineers Can Do Anything", "Engineers Color our World", "Women in Engineering" and "Engineering Technology" have prompted poster submissions from hundreds of students all over the country. This contest is effective because each student must research about engineering opportunities, and understand the contest theme in order to design a winning poster.

Students with a high proclivity for art will be excited to hear about a new program beginning at Rensselaer Polytech Institute (RPI). RPI's program is the Experimental Media and Performing Arts Center (EMPAC) which is a nexus of science, engineering and the arts. The research focus at EMPAC will provide students, researchers, artists, and audiences with opportunities for leading-edge science and engineering, performance technologies and art. The spaces and infrastructure are equally enabled for research that links sensory perception and human expression using computers.

All forms of communication are valuable and when using graphics or art to convey ideas, you may see the spark in a student's eye that wasn't there before.

Engineering Fairs

Career fairs attract companies that have full-time positions, part-time positions, internships and co-ops too. Students can talk to representatives from many companies about the field in general or about specific opportunities. It's a great opportunity to make contacts and learn from employers, and to discover many possibilities for employment.

In the K-12 world, students don't need jobs....yet. They are in the market for a college that meets their needs. Many colleges host events to recruit students from local high schools — these may be exhibition days with tours of the campus, showing projects done by student classes and teams, or they may be challenge events where high school student teams work on a particular problem and share their solutions. These may also be called "tour days" or something else, and they are great ways to get students motivated.

Many colleges are already doing this and, in fact, it is the number one type of event that I attend. Many colleges have invited me to give the keynote address to local junior high or high school students on their "Engineering Day." After the keynote, students are given a tour of the campus and speak to current engineering students. After the tour, which usually includes the labs and competition creations (such as concrete canoes, moon buggies or solar cars, tractors and/or airplanes) they complete some kind of engineering project. Engineering students have them complete projects within each discipline to give them a flavor of the different types of engineering. The pre-college students are kept busy until they get back on the bus to return to their respective schools.

These engineering fairs are excellent exposure to engineering and provide a positive and productive learning experience. Exposure to a college campus, engineering students and faculty are a winning combination. At the end of a day on campus, the pre-college students seem energized and excited about the possibilities.

This idea is also excellent for holding a "Girl Day" to recruit girls. The day is usually successful because it gives girls a chance to assess engineering as a career opportunity without the added pressure of trying to impress or deflect the boy across the room.

Chapter Five
Successful Curriculum, Programs and Projects

If you want to know what is going on in engineering education around the country, this list is for you. However, it's not all-inclusive. These are just a few of the more popular approaches to implementing engineering at the K-12 level. If you want more information about technology and engineering resources, visit the Boston Museum of Science's teacher-reviewed selection of the best standards-based technology and engineering curriculum resources for your classroom at *http://www.mos.org/tec.*

Pre-Engineering Curriculum and Projects

- **Engineering the Future**
  Engineering the Future (EtF): Science, Technology, and the Design Process (www.keypress.com/etf) is a laboratory course for the first year of high school science, created to help a broad spectrum of students. EtF is a full-year lab course organized around four projects, each of which is divided into several tasks. In the first project, students design solutions to problems that they find interesting. In the second, they design energy-efficient buildings to counter the problems associated with urban sprawl. In the third unit, they learn about thermal-fluid engines as they design and build toy putt-putt boats, and write patent applications for their innovations. The fourth project challenges the students to design electric circuits. Units two, three, and four illustrate how the same fundamental concepts of energy flow apply to thermal, fluid, and electrical systems. Kits are available for projects 3 and 4 and the entire course can be implemented on a modest budget.

- **Engineering is Elementary**
  This project develops curricular materials in engineering and technology education for children in grades K-5 (www.mos.org). Educator support includes lesson plans, assessment materials, and professional development programs that tie into other major content areas, including science and language arts. Each EiE unit focuses not only on engineering

and technology, but also on key standards in a specific science area. The program also conducts research into student and teacher learning of key concepts.

- **Project Lead the Way**
  Project Lead the Way (PLTW) is a non-profit organization that promotes engineering courses for middle (Gateway to Technology) and high school (Pathway to Engineering) students. The program formally partners with school districts, trains the instructors that will be teaching and implementing the curriculum, and acts as a bridge between educational institutions and private businesses. (www.pltw.org)

- **The Infinity Project**
  The Infinity Project is a national high school and early college math- and science-based engineering and technology education initiative that helps educators deliver a maximum of engineering exposure with a minimum of training, expense and time. Created to help students see the real value of math and science and its varied applications to high tech engineering. (www.infinity-project.org)

- **The Ford Partnership for Advanced Studies (Ford PAS)**
  Ford PAS is an academically rigorous, interdisciplinary curriculum and program that provides students with content knowledge and skills necessary for future success in such areas as business, economics, engineering, and technology. The inquiry and project-based program offers a series of modules that links learning in traditional academic subjects with the challenges students will face in post-secondary education, along with the expectations of the workplace they will face as adults. These links are forged through community-wide, cooperative efforts and partnerships that join local high schools, colleges and universities with businesses. (www.fordpas.org)

- **Materials World Modules**
  Materials World Modules focus on materials engineering – books, kits and training for middle and high school students.
  (www.materialsworldmodules.org)

- **Salvadori Center**
  The Salvadori Center focuses on improving children's content understanding and problem-solving skills by using project-based learning that focuses on the built environment. (www.salvadori.org)

- **Stuff That Works**
  Stuff That Works is technology curriculum for the elementary grades. City Technology introduces children across the country to the basics of design technology through curriculum materials, teacher resources, and professional development. (citytechnology.ccny.cuny.edu)

- **Children Designing and Engineering**
  Produced by the College of New Jersey, Children Designing and Engineering are teacher instructional guides that describe how to adapt activities for different populations, and provide hints for managing design-based learning (www.childrendesigning.org).

- **Teachengineering.org**
  Funded as part of the NSF-supported National Science Digital Library (NSDL) to provide educational resources for STEM (science, technology, engineering and mathematics) education. TeachEngineering.org is a searchable, web-based digital library collection populated with standards-based engineering curricula for use by K-12 teachers and engineering faculty. (www.teachengineering.org)

Exciting Initiatives

- **Design Squad**
  Design Squad is a reality-based competition show aimed at kids and people of all ages. Its goal is to get viewers excited about engineering and the design process. Over 13 episodes, eight high school contestants tackle engineering challenges for real world clients—from creating cardboard furniture projects for IKEA to designing a gravity bike (no pedals or cranks!) for Extreme Game champion Tom Whalen. The series has free engineering resources you can use in classrooms, after-school programs, and event settings to get middle school kids excited about engineering. (www.designsquad.org)

- **EPICS (Engineering Projects In Community Service)**
  EPICS teams undergraduates to design, build, and deploy real systems to solve engineering-based problems for local community service and education organizations. EPICS was founded at Purdue University in Fall 1995.

  EPICS programs are operating at 15 universities. Over 1,350 students participated on 140 teams in 2003-04. Peer teams at multiple EPICS sites are collaborating to address community problems of national scope. Purdue is headquarters for the National EPICS Program.

  EPICS appeals to people who want to make a difference in their community. At Purdue University there was an 11 to 15 percent improvement in the retention of female students who participated in this service learning-focused community, compared to female students in learning communities without a service-based approach.

  High school EPICS programs are being created in Indiana, California, Massachusetts, and New York. (http://epics-high.ecn.purdue.edu/)

- **Specialized High Schools**
  New York City, with its massive number of students has taken a different approach to ensure that each student has a chance of success. The specialized high schools of New York City are selective public high schools, established and run by the New York City Department of Education to serve the needs of academically and artistically gifted students. The Specialized High Schools Admissions Test (SHSAT) examination is required for admission to all the schools except LaGuardia, which requires an audition or portfolio for admission.

Each of the specialized high schools has its own features but most emphasize mathematics and science. They offer many electives and advanced placement courses, including enrichment courses in the humanities. Three schools, Brooklyn Technical High School, Staten Island Technical High School and the High School for Math, Science and Engineering at City College have an engineering focus.

The Texas Science, Technology, Engineering, and Math Academies are similar except they are non-selective and the majority of students are high-needs. The schools are small, approximately 100 students per grade, and encompass a personalized learning environment with high expectations. The academies serve grades 6-12 and are a mix of charter schools, traditional public schools, and schools created in partnership with an institute of higher education.

- **Magnet Schools**

  A magnet school is a public elementary, middle or high school that offers specialized courses or curricula. In the United States, where education is decentralized, some magnet schools are established by school districts and draw only from the district, while others (such as Governor's Schools in Virginia) are set up by state governments and may draw students from multiple districts. Some magnet programs are within comprehensive schools, as is the case with several "schools within a school." In large urban areas, several magnet schools with different specializations may be combined into a single "center," such as Skyline High School in Dallas, Texas and Advanced Technologies Academy in Las Vegas, NV.

  Districts started developing magnet schools to draw students to specialized schools all across their districts. Each magnet school has a specialized curriculum and students choose a school based on their interests. Competitive entrance processes have been implemented to encourage good grades and behavior from students who wish to be accepted into the school of their choice.

Learning Laboratories

Many companies also work to bring engineering education into focus. A few of the more prominent and proven examples include:

- **Pitsco — Academy of Engineering**
  Pitsco's Academy of Engineering is a continuum of hands-on curriculum that provides engaging, standards based content in STEM. Three years of teacher-led, student responsible projects are presented with real-world engineering fields and contexts such as aeronautical, mechanical, civil, rocketry, robotic, automotive and green engineering. Each engineering context spans nine weeks (one quarter) of content and experiences.

- **PCS Edventures — Academy of Engineering (AOE)**
  The AOE is a mobile engineering laboratory that combines hands-on activities with either Fischertechnik® or LEGO® Manipulatives to teach students science, technology, engineering, math, architecture, communications, robotics and more. It's a STEM solution with hundreds of hours of course work and activities. The program also includes online teacher training, student assessment and support, and a virtual online community that includes quarterly engineering challenges and at-home extensions (www.edventures.com/AOE).

- **Electronic Circuits for the Evil Genius**
  Provides about 100 hours of electronic hardware training for ages 13+. Meets Canadian standards for high school electronics classes (www.elxevilgenius.com).

Beyond K-12

Once you find curriculum or programs that your students enjoy and they are ready to go to engineering school, you'll be happy to know that engineering schools and colleges have become very creative in their quest to attract the best students.

- Some are increasing the degrees offered. Many have started programs in biomedical engineering, nanotechnology, sustainable engineering and other hot areas that are of interest to students.
- Some schools have engineering residence halls.

- Some have international programs such as Engineers Without Borders.
- Some have engineering sororities and engineering student organizations for all disciplines and interests.
- Some offer a large variety of scholarships.
- Some schools don't allow students to select a major until their sophomore or junior year.
- Some schools offer design courses for freshmen.
- Some offer summer camps for high school students and competitions for undergraduates.
- Many have Women in Engineering as well as Minorities in Engineering programs.
- Many have co-op programs and internships.

The following programs are national and serve undergraduate engineering students. It's important to know about these programs and present some of the opportunities available when helping students make the decision to go to engineering school.

- **The Bachelor of Art in Engineering**

A 1995 NSF survey found that only 38 percent of those in the U.S. workforce with a degree in engineering actually work as engineers. Another 4 percent say they work in a related science field; and an additional 48 percent aren't considered engineers or scientists, but say their work is related to engineering. With an increase in technology in our daily lives, many people now view engineering as a marketable skill. Many newly emerging non-technical fields need the logical problem-solving abilities of engineers. The Bachelor of Arts (BA) in Engineering is considered to be the liberal arts degree for a technologically driven society. Most BA programs in engineering are designed to appeal to a broader segment of the population than the BS degree. Graduates of these programs may find themselves working in public policy, management, business or any other sector of society that can benefit from a thorough grounding in the engineering process.

- **Undergraduate Engineering Education Departments**
  Purdue University established the first engineering education department in 2004. The Virginia Tech Department of Engineering Education was established later that year. Utah State University and others are following in their footsteps. These departments are beginning to find out how students learn engineering and to research the connections to pre-engineering pedagogies and methodology in the K–12 community.

- **Engineers Without Borders USA (EWB-USA)**
  EWB-USA is a non-profit humanitarian organization that partners with developing communities worldwide to improve their quality of life. The partners implement sustainable engineering projects while training internationally responsible engineers and engineering students. EWB-USA is a good hook for inviting women into engineering as many chapters are comprised of 50 percent women. EWB-USA appeals to women who want to make a difference globally.

- **Design that Matters (DtM)**
  DtM is a nonprofit organization that creates new products to allow social enterprises in developing countries to offer improved services and scale more quickly. Hundreds of volunteers in academia and industry donate their skills and expertise to the creation of breakthrough products for communities in need.

- **Graduate Certificate in Secondary Teaching for Engineers**
  For the engineer, engineering student or professional that wants to become a teacher, some universities such as Boise State University and Oregon State University offer a Graduate Certificate program. This can often be completed with just one additional year beyond the Bachelor of Science (BS) degree. Students with a bachelor's degree in an engineering field can apply to the graduate certificate program either after years of

experience as an engineer, after completion of their bachelor's degree, or even in the final year of their BS program.

As you can see, there are many different programs and roads to becoming an engineer. What students want, dream about and need all determine the path they take. There is something for everyone because engineering can serve as a fantastic launching pad to an assortment of careers and world wide opportunities.

Chapter Six
14 Exciting Hooks That Engage Student Interest

Part of the problem with getting students to recognize the great part of engineering is that they aren't making the connections between their favorite things and engineering. This chapter is dedicated to helping you uncover for your students the possibilities of an engineering career by showing some of the exciting and colorful opportunities available. It is impossible to cover them all — there are just too many. This compilation is simply my perspective and experience of what's "really interesting" to K-12 students, and should only serve as a starting point in locating engineering careers. Many of the career tracks you'll find are multidisciplinary and, therefore, usually more attractive to today's youth.

Top 14 Hooks for Motivating Students

1. Sports Equipment Design

Sports engineering is an excellent way to impact athletes, sports and businesses around the world. Engineers in this field are some of the most dynamic, innovative and creative engineers on the planet. Not only is this industry full of diverse, fun and intriguing opportunities, but most of the engineers working in the sporting goods industry became engineers because they love sports and wanted to either increase their performance or enhance the sport overall (Baine, 2004). A company that wants to design a new golf club for Tiger Woods would prefer to hire an engineer who plays golf. A company that is designing new high-performance mountain bikes would prefer to hire an engineer who has a keen interest in bike design or racing bicycles. The industry offers some awesome careers for athletically inclined engineers!

For example, chemical and materials engineers will find a wealth of employment in the athletic shoe industry. Constantly on the lookout for new materials for soles and outer coverings, shoe manufacturers are in ongoing competition for the best cushioning, lightest overall design, most comfortable and best traction products. Finding new materials that add "breathability" for the long distance runner, springiness for basketball players, increased traction for skateboarders, flexibility and grip for wrestlers, more cushioning for long jumpers and/or strength and comfort for skeleton racers can make the shoe industry a challenging and rewarding field for an athletically minded engineer.

On a pair of athletic shoes:

- Mechanical engineers may design systems for manufacturing, motion analysis or impact testing and be involved in building and/or testing prototypes.
- Biomedical engineers may design systems for motion analysis and biomechanical analysis of injuries, stress patterns, or kinesthetic optimization.
- Chemical, materials or textile engineers may develop or design new soles, fabrics or other materials for shoes.
- Manufacturing engineers may design systems or processes for manufacturing shoes more efficiently.
- Computer engineers may design software or hardware to aid in pressure or impact detection analysis, manufacturing processes or information systems.
- Industrial engineers may maintain the bill of materials and routing information, cost standards and recommend pricing for new products. Or, they may be involved in learning about and training on manufacturing techniques.

Because sports in an integral part of many teenagers' lives, have them brainstorm and research who and what it takes to manufacture their favorite piece of equipment.

2. Animatronic and Animation Engineering (Lucasfilm, Movie Studios and Pixar)

For many movies such as *Jurassic Park* and *King Kong*, full-scale animatronic (animatronics is the electronic technology used to animate motorized puppets or characters) characters are made along with animated computer versions of the same creature or dinosaur. Engineers may develop new electronics to make an animatronic character function. Or, engineers in this industry may create or program the digital version of the same character. Whatever the case, engineers make everything 'stand up.'

3. Roller Coasters and Themed Attractions

Creativity and good communication skills are essential in this field. The conceptual designer might say to the engineer, "I want the guests to smell bananas on King Kong's breath." The engineers will work from that starting point and determine the feasibility of the effect. To determine feasibility, they have to approach the problem from the perspective of safety, reliability, guest impact, and budget by asking questions such as, "Will the people be able to smell the bananas? Is it safe for King Kong to get close enough so that people can smell the bananas? How much smell is too much? How much is not enough? How much money will it cost to produce this effect? What effect will smelling the bananas have on the people? Will it affect people that are allergic to bananas? Do we post a warning sign about banana allergies at the beginning of the ride?" and more. Merging art and science to create amazing theme parks is different from engineering in any other medium. Animatronic engineers face constant unusual challenges that may or may not be easy to solve.

In 2005, 253 million people visited amusement parks. As long as people want to be hurled around roller coaster tracks at incredible speeds, terrified and exhilarated but completely safe, too, there will be jobs for roller coaster engineers. Roller coaster enthusiasts are constantly in search of steeper hills, more G's, tighter rolls, and more airtime.

According to a 2008 roller coaster census, there were 2076 operating coasters worldwide. Roller coasters are a prominent feature in many theme parks. Today's coasters are built bigger and faster than ever. Parks now have 30 or more different types of coasters to choose from and are adding them by the handful. During an interview with Park World Magazine, Jim Seay, president of Premier Rides, observed that "Advancements in technology are allowing manufacturers to build new and unique types of coasters which give theme parks unlimited choices when shopping for the next great marketable attraction. Theme parks are buying coasters because it is smart business. It has been proven time and time again that new coasters quickly pay for themselves and bolster annual attendance." When a park adds a new coaster, the average attendance increases by 20-25 percent. That increase is a strong statement for the business side of introducing new roller coasters.

Imagineers are not the only type of engineer that a theme park such as Disney, Busch Gardens, or Universal Studios requires. Theme parks often need industrial engineers, ride and show engineers, project engineers, architectural engineers, electrical engineers, computer engineers, mechanical engineers, structural engineers, and audio/video engineers (Baine, 2007). Imagineers may design the face of what the public sees, but the work of other engineers supports the business of theme parks in areas such as crowd control, research and development, training, and equipment maintenance.

Other industries that hire engineers to design, build, conceive, produce or orchestrate entertaining environments include themed retail stores, themed restaurants, sports facilities, museums, aquariums, zoos, casinos, cinemas, expos, recreational facilities, coliseums, and planetariums. In this field, electrical and computer engineering are the most common degrees. Mechanical engineers may design sets or animatronics/hydraulic systems and materials engineers may develop new coverings to make animatronic characters more life like.

4. Eating

Eating is not only necessary for survival but sharing a meal is also a way to get to know someone, have fun, celebrate an occasion, experience a different culture and to say, "I love you." Food is the center of attention at these events and food engineers are involved in all aspects of food preparation and processing. They influence the packaging, storage and distribution systems of foods as varied as candy bars and frozen dinners. Because new products and environmentally friendly food processing equipment need to be developed, engineers are in demand within the food industry.

Food engineering pertains to the properties and characteristics of foods that affect their processing. Food engineering requires an understanding of the chemical, biochemical, microbiological, and physical characteristics of food. There will be a shortage of food engineers as long as society demands that engineers develop lower fat, lower salt, lower cholesterol, or nutrient-packed foods for their diets. Food engineering is a branch of chemical engineering but students that want to work in the food industry can also major in biomedical, mechanical or manufacturing engineering.

5. Music

Music, audio, electrical, and computer engineers are a major component of a team of people that create iPods, digital instruments, microphones, speakers, headphones, Internet music sites such as iTunes, theatre sound, and live concert sound. They work in music studios, for video game companies and for other software companies that need audio support. They focus on revolutionizing the music and sound industries and finding new ways to listen to, create, store and preserve music (i.e. music on iPods, cell phones, etc.) (Baine, 2007).

There are many approaches to combining engineering, technology and music. In general, a career in music engineering or technology requires that a student be musically inclined as well as technical and creative. If they love music, like to work on computers, are fascinated by electronics and mechanics, or have a love for gadgets, combining music with engineering or technology may be the solution to a satisfying career. Not only can it lead to a successful career contributing to the newest releases on the charts, but it can also lead to success creating instruments or changing the way we listen to music. The good news is that this field is wide open with plenty of opportunities for a hard working, ambitious person.

Students can get a formal engineering degree, a certificate or degree in music engineering technology, or they can be self-taught. They can go to school for four or more years, two years, one year, one month, one week or one hour. Whatever their passion and ambition, there is an accessible career path. The most common majors for music engineers are audio, electrical, computer and software engineering.

6. People Engineering

Biomedical engineering is, in a very real sense, people engineering. The objective of biomedical engineering is to enhance health care by solving complex medical problems using engineering principles. Those who specialize in this field want to serve the public, work with health care professionals, and interact with living systems. It is a broad field that allows a large choice of sub-specialties.

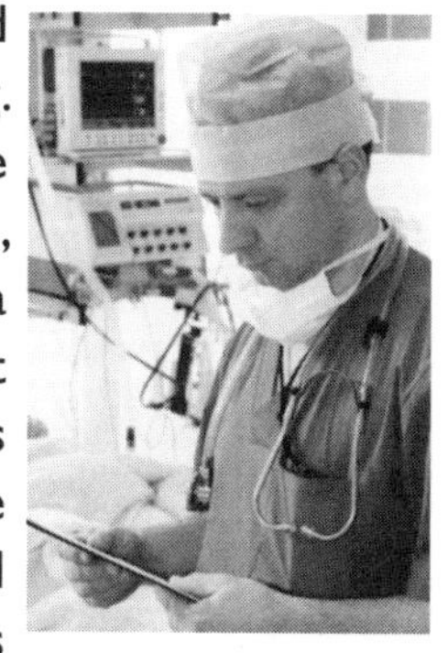

Biomedical engineering is interdisciplinary and newer than most other types of engineering. Because of its interdisciplinary nature, it is possible to get a degree in mechanical, electrical, chemical, or materials engineering and still work as a biomedical engineer. In fact, 30 years ago most biomedical engineers were trained in other forms of engineering because of the lack of available programs. Students who major in mechanical engineering can create prosthetics. Students who major in electrical engineering can design hospital equipment or laser treatment for cataracts; and students who major in chemical engineering can design drug delivery systems. Many students say they chose biomedical engineering because it is people-oriented.

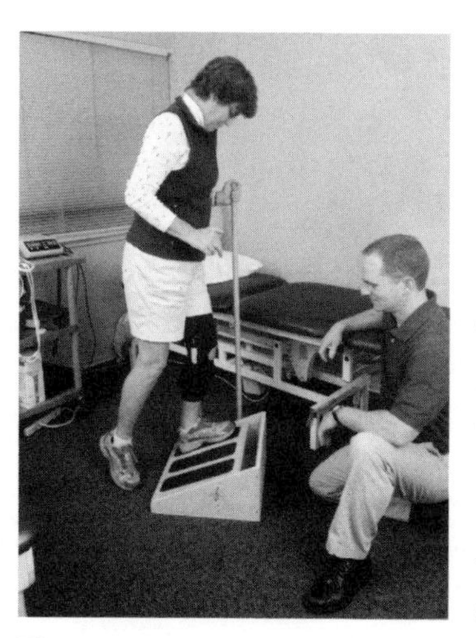

With new advances in technology developing daily, biomedical engineering is growing and evolving quickly. Biomedical engineers design instruments, software, and medical devices to help physicians, nurses, veterinarians, therapists and technicians solve complex medical problems and enhance the quality of medical care.

Imagine designing a medical device that allows blind people to see or allows a heart to beat rhythmically. The pacemaker was invented by biomedical engineers who literally gave recipients the ability to perform physical activities, such as

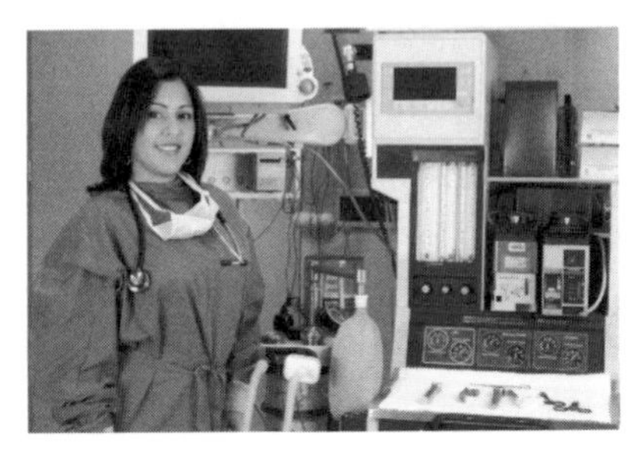

climbing a flight of stairs or walking around the block that they were previously unable to do. A few years ago, four engineering students at Northwestern built a prosthetic device that allowed a burn victim who had lost a hand in a fire to play tennis again.

Women seem naturally drawn to biomedical engineering. According to the American Society for Engineering Education, biomedical engineering leads all engineering disciplines in the percentage of degrees awarded to women at all levels — bachelor's, master's and doctoral. Forty-five percent of all Bachelor of Science degrees in biomedical engineering in 2004 went to women.

Many women grow up interested in engineering but want to do something that will affect people on a personal level. Biomedical engineering can have a direct impact on people's lives and on their health, as opposed to making a better gear for a car or designing it to go faster.

7. Space Engineering

Assuming that astronauts, spacewalks and rockets already enamor students, the greatest thing about motivating students to pursue engineering or technology in this arena is that they can be almost any type of engineer. When I was a kid, I thought that I had to be an aerospace engineer to become an astronaut and get a job at NASA. I had no interest in aerospace so I never even entertained the idea of being a part of the space program.

For students today, I'm happy to say that I had bad information. NASA employs nine engineers for every scientist (NASA, 2007). They hire biomedical engineers to make space suits, chemical engineers to help with life support systems, mechanical engineers to work on almost everything, electrical engineers to work on control systems, etc. There are endless opportunities for students in this field and they don't have to narrow their interests to succeed. According to Reliable Robots on the Futures Channel, some engineers might spend their days "fine tuning a set of million dollar micro-cameras so the rovers can 'see' better while exploring miles of Martian terrain". Or, they might be "designing tele-operated mini-rovers in an office that looks more like a high tech R&D lab at a toy company than a NASA research facility".

Dr. Sally Ride gives great advice to students. She says, "The most important steps that I followed were studying math and science in school. I think the advice that I would give to any kids who want to be astronauts is to make sure that they realize that NASA is looking for people with a whole variety of backgrounds: they are looking for medical doctors, microbiologists, geologists, physicists, electrical engineers. So find something that you really like and then pursue it as far as you can and NASA is apt to be interested in that profession."

NASA's engineers brought us inventions such as the hand held vacuum cleaner, the firefighter breathing apparatus, safer runways, storm warning systems, better sunglasses, car crash technology, freeze dried meals, baby food, improved air quality, artificial limbs and much, much more. Engineers at NASA help our daily life in ways that often go unrecognized.

The NASA web site is full of information to help you engage students. Watch videos of the shuttle launch and even see it from a camera on the solid rocket boosters. There are numerous pictures and classroom activities for every age. It's easy to get information and students involved by visiting *www.nasa.gov*.

8. Animal Engineering

Engineering medical equipment and life support systems for animals is a big field for engineers. In 2007, NBC News reported that Americans spent $180 billion on their pets. Pet care is a booming industry and a fantastic field for engineers. Many girls are also drawn to the field with the hope of helping both small and large animals. Have students imagine that they are responsible for building a new exhibit for an endangered or rare tiger. The animal has to be happy in its new home, they need to make sure that the animal cannot escape from its habitat and that the guests have a nice viewing area without encroaching on the Tiger's personal space. Have the students determine what kind of information they need (how high does a tiger jump, what's the perfect climate, do they like other tigers, etc.) and design an exhibit that will make the tiger happy.

Zoo design is a tricky business. It's interesting to learn about penguins, hippos, rhinos, giraffes, etc. The work is similar to designing building systems but the equipment used in a zoo is unusual. Engineers usually have to go to special training for it, and it's important for them to keep up with the changes in the industry. Not only do engineers have to research and design the perfect type of environment for each animal, they also have to keep the zoo visitors happy by creating something that is aesthetically pleasing. Zoos spend money to obtain new animals and build exhibits in order to attract more guests and stay in business.

Architects usually design the face/aesthetics of the zoo but engineers are required to keep the animals alive and healthy. Engineers who work for zoos are called life support engineers and are usually educated in civil, chemical, biomedical, environmental and mechanical engineering.

9. Computers and Connectivity

Computer engineering is another very interesting and fast growing field. Computers are in automobiles, microwaves, iPods, watches, telephones, mobile devices, video games, and much more. It is very similar to electrical engineering, except that computer engineers work exclusively with computers and computer systems or equipment.

- Computer engineers develop hardware technologies such as computer architecture and the architecture of networks and systems that link computers together.
- Software engineers deal with applications such as artificial intelligence and operating systems that run all computers and related systems.
- Network engineers research, design and install the architecture and networks that computer engineers develop.

These engineers may research, design, develop, test, manufacture, or install computer systems, networks, circuit boards, integrated circuits (computer chips), operating systems, software or peripheral equipment such as keyboards, mice, printers, speakers, or microphones. They may also plan computer layouts, or research future applications or environments for computers.

Computer engineers may be at work right now trying to figure out:

- How to embed computers into shoes that will grow more spring if you are walking or running quickly;
- Prescription eye glasses that can eliminate bi- or trifocals by detecting eye strain; or
- Ovens that can double as freezers and be programmed to turn on, defrost, and then cook your dinner.

The computer revolution has created countless jobs in every field of technology. Computer engineers, computer engineering technologists, computer scientists, and information technologists are all in high demand. Information technology has created abundant job opportunities for the talented and highly skilled computer engineer.

10. Doctors, lawyers, CEO's, writers, teachers, politicians, entrepreneurs, inventors

Exposure to engineering in secondary school is wonderful preparation for a wide variety of roles within society. Diverse and plentiful opportunities exist for the educated non-mainstream engineer with a good understanding of scientific and technical subjects. According to a National Science Foundation study in 1998, out of 2.2 million degreed engineers, one million were not working primarily as engineers. Students with a degree in engineering and an interest in medicine, law, finance, writing, teaching and politics typically have very successful careers. There are many opportunities for the engineer in search of alternatives to traditional industry, because engineers can do anything.

- According to the American Medical Association, students with bachelor's degrees in biomedical engineering have a higher acceptance rate into medical school than students with any other undergraduate degree.
- If a student wants to become an attorney specializing in environmental law, a good way to start would be with an undergraduate degree in environmental engineering; a mechanical engineering degree would be a good foundation for someone who wants to become a patent attorney.
- Students with an interest in finance may find their way to Wall Street to become financial engineers.
- Students with an interest in writing or teaching become technical writers or teachers.
- Students with an interest in politics can become public policy experts. Three engineering students even went on to become

president of the United States! New opportunities also abound for engineers in the House or Representatives and the Senate. These engineers set public policy and inform the government on how they may be able to create better and more responsible public policies.

These professions require analytical, integrative, and problem-solving abilities, all of which are part of an engineering education. Thus, engineering offers an ideal undergraduate education for living and working in today's technologically dependent society.

Engineering also lends itself nicely to entrepreneurial types. Engineers of all types have a wildly successful history of ventures. Great at creative problem-solving, some engineers seem to have innate knowledge about how to be more efficient, make things easier to use, create new products, or save money. Someone (probably an engineer) invented or improved most of the products or items that you can see, touch, taste, smell and hear.

Other engineers form construction, environmental, or computer consulting firms because their knowledge is in demand. Engineering schools help students develop entrepreneurial skills such as dealing with venture capitalists, writing business plans, and understanding the business processes of a company. Students have to understand not only what it takes to technically create an invention but the financial and marketing aspects of their invention as well.

11. Sales, testing, systems, design, quality, etc

Students who enjoy working with other people and traveling may become sales or field service engineers. Students who enjoy the big picture may become the systems engineers who put all the pieces together. Creative students who constantly have new ideas about everything may enjoy working as design engineers. Students who enjoy breaking things, conducting experiments or working in laboratories may enjoy working as test engineers.

12. Feeding the World

One in seven people around the world is suffering from hunger. By 2025, 2.8 billion people will be facing fresh water shortages. Our population places great demand on our limited natural resources. For the student that wants to be a part of the solution, biological and agricultural engineering is a great choice. Biological and agricultural engineers (B&AEs) work to ensure that we have the necessities of life: safe and plentiful food to eat, pure water to drink, clean fuel and energy sources and a safe, healthy environment in which to live.

B&AEs devise practical, efficient solutions for producing, storing, transporting, processing and packaging agricultural products. They solve problems related to systems, processes, and machines that interact with humans, plants, animals, microorganisms and biological materials. They develop solutions for responsible, innovative uses of agricultural products, byproducts and wastes, and of our natural resources of soil, water, air, and energy, etc. And they do all this with a constant eye toward improved protection of people, animals, and the environment.

B&AEs may find new uses for agricultural products, byproducts, and waste, develop industrial air filters embedded with microorganisms that help reduce air pollution; determine improved methods of soil erosion control; study animal behavior to develop more humane housing environments; and/or develop renewable energy sources from grain oils.

Students would enjoy a career in Biological and Agricultural Engineering if they want to work with people, help make a more sustainable future, they have a "green" focus, they enjoy working with plants or animals and they enjoy good food.

13. Energy Engineering

Each of us uses energy everyday. Not only do we use energy to walk, talk, play sports, and function (all of which use calories) but we also use energy to power our cars, toast our bread, watch TV, on so on. Energy is everywhere and there are multiple forms of it. Energy can be kinetic (electrical, thermal, geothermal, nuclear, light, motion, water or sound energy) or potential (chemical, nuclear or stored energy).

When energy is renewable it means that it can be re-used. For example, energy from the sun can provide power when it hits a solar panel. The sun's energy is renewable because the energy from the sun is still available even after you use it. If the same power was being provided by natural gas, once you use the gas, it is gone forever because natural gas is a non-renewable energy source.

Engineers are hard at work designing ways we can use renewable energy. Engineers in this industry are designing engines that run cleaner for improved efficiency as well as developing electric and hybrid vehicle batteries and systems. Other engineers are working to improve the efficiency of wind, water and solar power. Still other engineers are exploring the potential of future technologies utilizing wind, solar, geothermal, biofuel and wave energy sources. These need to be cultivated, expanded and implemented, as well as meeting the increased demand for greener buildings and transportation systems.

Jobs in the green collar sector — such as solar panel and turbine manufacturing, installation, sales, research, and design are in high demand. Renewable energy technologies diversify our energy supply, reduce our dependence on imported fuels, improve air quality, offset greenhouse gas emissions and stimulate the economy. There are currently no degree programs for sustainable engineering although programs in renewable energy systems, solar power and wind energy are available.

14. Save the World Engineering!

Going green is "hot" right now. Programs all over the country are showing increased enrollment in anything that involves sustainable engineering. Students who have an interest in saving the world will be interested in sustainable engineering since it focuses on the development of tools, knowledge and processes for a sustainable future. Areas of work include preventing pollution, assessing the environmental impact of everything, recycling methods, and renewable energy. These earth-friendly professionals concern themselves with preventing and fixing problems caused by the previous industrialization era. They concentrate on delivering better environmental conditions for the public through knowledge, research, a caring attitude, and common sense.

One of the most rewarding aspects of being a sustainable engineer is that because there is such a large need for sustainability on every level, enthusiasts can make a difference right away — from the first day on the job right through the rest of their career. This fast-growing field offers a challenging and satisfying chance to protect the health and safety of people and our environment.

Those students who want to work with people, help make a more sustainable future, save the planet, protect the rainforest, save the polar bears, or capture and store renewable energy sources will find sustainable engineering a booming field. Jobs will be created because our planet needs environmentally sustainable solutions to support population growth and preserve our limited natural resources.

Companies that employ sustainable engineers include civil engineering firms, NASA, The Department of Energy, The Department of Agriculture, and the Environmental Protection Agency. Many large firms also employ sustainable engineers to reduce their environmental footprint. Some of these firms build, design or inspect green energy systems and others develop and administer the regulations that protect our health, safety, and the environment.

The most common majors that work in the field of sustainable engineering are agricultural and biological engineers, chemical engineers, civil engineers and environmental engineers. However, many other fields of engineering such as mechanical, materials, electrical, and Heating, Ventilating and Air-Conditioning (HVAC) engineers also work on projects in this field. For example, a mechanical engineer might help design wind turbines and an electrical engineer may work for a company that designs solar panels. As this field continues to grow, so will the opportunities for many different types of engineers.

We all want to make a difference. Fortunately, engineering is one of the best ways in which to achieve it because the work of engineers impacts millions of people everyday.

Chapter Seven
For the Parents of Budding Engineers

This chapter is for the parents reading this book. Being involved as a parent gets it's own chapter because if parents don't understand what engineering is and are unsure about the opportunities available to their son or daughter, the student will not get the support at home that is critical to his or her success.

To help in your child's career development, be involved. Talk to him or her about favorite activities and interests, and dreams for the future. Be sure that you examine your own views about engineering so that you don't unintentionally reinforce a negative attitude (i.e. saying nerd instead of geek, or admitting that you hated math or, if it's your daughter, making her think engineering is only for boys). And lastly, have high expectations for your child. Expect him or her to be interested in and do well in engineering, and share how you feel.

To interest more students in engineering, both educational institutions and households must make a concentrated effort to breed an atmosphere of encouragement. We must convince students that technical careers are not just attainable but fulfilling.

Ten Strategies to Nurturing an Engineer

1. **Your child can do it!**
   Remember that math and science grades are not always good indicators of success in engineering school. "I love science" is often a better indicator. My son claims that math is his favorite subject. However, he only has a C in the class because he forgets to turn in his homework. Grades in his case are a poor indicator of his ability and potential.

2. **Don't pass on bad math attitudes**
Engineering is not all math. It's just one of the tools in the engineer's box. Show your child that math and science are fun by making real world connections. My daughter became very skilled at math because when we went shopping for clothes and the sale price was 20 percent off, she knew she wouldn't get that beautiful jacket unless she could tell me the correct price.

3. **Help your child explore careers**
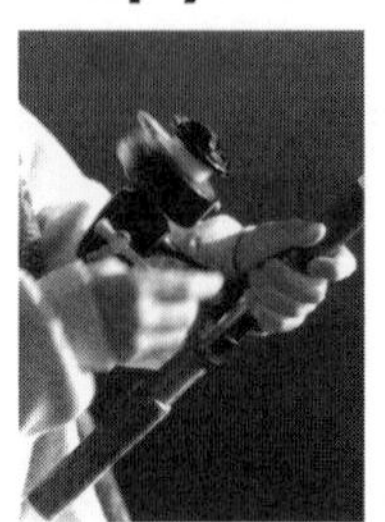
I talked to an engineer who told me he loved to fish as a kid. Every chance he got he was out fishing. Wouldn't it be great if your child found the perfect job within his or her favorite hobby? The guy in the fishing story is now the head fishing reel engineer for Pure Fishing, Inc. There are countless stories about engineers finding their dream jobs through their hobbies.

4. **Enroll your son or daughter in an engineering camp**
Camps are a great way to expose your son or daughter to engineering. You can find more information about summer camps on page 70.

5. **Promote after-school activities**
After-school programs in robotics or math are available at many locations. The best place to search for a quality after-school program is your child's school. To find more programs you can also contact your local Child Care Resource and Referral (CCR&R) agency. It is a community agency that can help you choose high-quality, after-school care that meets local regulations and standards and that best meets your needs. The National Child Care Information and Technical Assistance Center (NCCIC), a service of the Child Care Bureau, hosts a website that provides the contact information for all State CCR&R agencies at *http://nccic.acf.hhs.gov/statedata/dirs/display.cfm?title=ccrr.*

6. **Provide subtle communication**
If your kids are typical teenagers, sometimes it's very hard to talk to them about career opportunities. If I ask my children to look at a book or catalog, they find a million reasons to ignore my request. A successful

strategy in my house is to very quietly leave college catalogs or career books lying around the house. Make sure they are visible but not too obvious. After a few days or weeks, you may notice that the book or catalog has been moved.

7. **Supply direct communication**
Many students form their attitudes about careers as a result of their interactions with family members. This can be used to your advantage by inviting to dinner any engineers or people in the field of technology. Encouraging that person to talk about his of her career – how he or she got into it and why it's satisfying. This can be a natural springboard for your child's questions and exploration.

8. **Take educational vacations**
When you travel around the country or even in your local area, there are many sights that will help your family learn about engineering. Places such as Hoover Dam, the National Inventors Hall of Fame, Thomas Edison's Birthplace, Museums of Ceramics or Aeronautics, roller coasters, etc. can all be educational and fun too. For sights in your area or to help you plan a road-trip, visit *www.engineeringsights.org*

9. **Visit the websites of engineering colleges**
Sit down with your child and check out the websites for your local colleges of engineering. Find out what is going on in your local area and look for ways to be involved. Make notes of what each school offers and especially about what seems exciting to your child. Make sure they know how to look for important information such as scholarships and entrance requirements. You can never do this too soon.

10. **Find a mentor**
Mentoring is successful because it's a one-on-one learning experience that can be so much more than a technical learning experience. Mentors can help students learn approaches into competitive industries, help them network, introduce them to key players, teach them how to listen, and help them evaluate solutions to problems. Mentoring is a

part of being successful in any industry but especially for careers that are competitive.

Chapter Eight
Recommendations and Conclusions

If you were to assess your problem-solving capabilities right now, what would you say about them? How do they help you? Before I began engineering school, my serious problem-solving technique consisted of a pros and cons chart. And then, I would still choose the answer that was emotionally stronger. An engineering education teaches you how to think through a problem in order to solve it. These mental agility skills will help you solve problems for the rest of your life.

The fascinating aspect about problem-solving in engineering is that there is almost never a "right" answer. Engineers access several different approaches to solve a problem, and then it is up to them to show everyone how their solution meets the needs of the client — or in the case of the Gilbreths — the family. The movie *Cheaper by the Dozen*, starring Steve Martin, was the real-life story of the Gilbreths — two industrial engineers trying to raise 12 children. The father was an efficiency expert who ran his household like a factory, and Mrs. Gilbreth applied her professional skills to raising and educating all 12 kids.

In elementary and middle school, it became very clear to me that I was not good at multiple choice tests. I would read the problem, peruse the answers A, B, C or D and become frustrated because I thought there should be options for E and F. Although irritating, this led to an appreciation of math, a subject where you could get partial credit. I celebrated the fact that if I took a cookbook approach to most math problems, remembered to account for details such as plus or minus signs, I could be reasonably successful.

Engineering was even better than that! In engineering school I learned that there are many solutions to a problem. Hurray! The challenge was picking the best one based on who or what it served. I saw this as a way to connect with people and eventually become the hero who made life easier or better for an individual or a company. This was life changing information because no one ever said that about the work of engineers. The more I learned, the more amazing the profession of engineering became. I couldn't figure out why this information was left out of all the books. For me, it was the best motivator of all.

Motivating students to pursue engineering is no small task. It requires constant ingenuity because every student is different. But, the point is to keep trying. There is something for everyone in engineering and the nation is sorely in need of a diversity of talent.

Recommendations

Suppose that a student approaches you with an interest in engineering. Let's also suppose that the student doesn't know much about the types of engineering available. What do you tell him or her?

The best approach is to whip out a purple cable tie and ask about their interests. The cable tie is perfect because you want to help him or her organize their interests into a colorful career path:

- You could recommend a viewing of the seven-minute presentations in the *Engineering Power Pack: A Career Presentation Bundle* or surf the Internet for information.
- You could recommend finding an engineering-related camp, club, competition or group to join. This can include social network engineering groups on MySpace or Facebook.
- Recommend touring engineering colleges or visit college of engineering websites to become familiar with different schools and approaches to learning.
- Recommend taking as many math, science, technology, computer, foreign language and art classes as possible.
- Recommend working on developing excellent communication and teamwork skills.
- Review the *50 Reasons to Teach Engineering* (pages 46-48) and see if it sparks any interest in the student.
- You could link the student's interests to engineering using the following table (the table is very generalized, you my need to make creative associations) and recommend seeking more information on careers related to the student's interests.

Student Interests	**Encourage Students to Explore:**
Listening to music	Electrical, Audio, Computer and Software Engineering
Going to the movies	Electrical, Mechanical, Computer and Software Engineering
Reading a book	All forms of engineering need technical writers or editors.
Socializing with friends	All forms of engineering develop communication tools.
Playing sports	All types of engineers develop sports equipment.
Riding roller coasters	Civil, Structural and Mechanical are the most common.
IMing friends	Electrical, Mechanical, Computer Telecommunications and Software Engineering
Talking on the phone	Electrical, Mechanical, Computer Telecommunications and Software Engineering
Eating	Chemical, Food, Biomedical, Mechanical and Manufacturing Engineering
Going to the beach	Ocean, Civil, and Environmental Engineering
Fishing	Mechanical, Materials, Electrical and Computer Engineering

Student Interests	Encourage Students to Explore:
Skiing	Mechanical, Manufacturing, Biomedical, Materials Engineering
Cooking	Chemical and Food Engineering
Camping	Civil and Environmental Engineering
Watching TV	Electrical, Audio, Computer, Software and Manufacturing Engineering
Playing an instrument	Electrical, Audio, Computer and Software Engineering
Surfing the Internet	Electrical, Computer and Software Engineering
Inventing stuff	All kinds of engineers invent stuff
Going shopping	Industrial Engineers work for Macy's, Old Navy, and other large clothing chains.
Playing video games	Electrical, Computer and Software Engineering.

An Achievable Goal

A fundamental problem with interesting students in engineering is that many don't understand what engineering is, what engineers do, how it impacts lives and how it can be a rewarding career. An excellent goal is to ensure that every student entering high school has or will be exposed to engineering, and has the resources available (mentors, books, videos, knowledgeable teachers and counselors, etc.) to make an informed choice on whether or not to pursue engineering.

Winning Attributes

As a society we need to accept math and science as tools to understand the world and solve problems. To motivate and inspire students, we as educators need to provide exposure, awareness, meaningful interactions, curricula, content, relevance to success, "aha" moments, and sustained engagement leading to a student's pursuit of an engineering degree. Positive attitudes and support are crucial in the classroom. Keep reminding students that engineering is a great way to make a huge contribution to society because engineers can do anything!

Appendix

References

- Achieve, Inc. 2005. National Data Profile. Available online at: *http://www.achieve.org/node/708*
- Allen, T.J. 1984. "Distinguishing Engineers From Scientists." In Managing Professionals in Innovative Organizations. R. Katz, ed. Cambridge, MA: Ballinger Publishing, 3-18.
- American Association for the Advancement of Science (AAAS). Project 2061. 1993. Benchmarks for Science Literacy. New York, NY: Oxford University Press.
- American Association for the Advancement of Science (AAAS). Project 2061. 1989. Science for All Americans. New York, NY: Oxford University Press.
- AP (Associated Press). AOL Instant Messaging Trends Survey Reveals Popularity of Mobile Instant Messaging. Available online at *http://www.businesswire.com/portal/site/google/index.jsp?ndmViewId=news_view&newsId=20071115005196&newsLang=en* Retrieved September 27, 2008
- Baine, Celeste. 2004. *Is There an Engineer Inside You? : A Comprehensive Guide to Career Decisions in Engineering.* Belmont, CA. Professional Publications, Inc.
- Baine, Celeste. 2007. *The Fantastical Engineer: A Thrillseeker's Guide to Careers in Theme Park Engineering.* Springfield, OR. Bonamy Publishing.
- Baine, Celeste. 2004. *High Tech Hot Shots: Careers in Sports Engineering.* Alexandria, VA. National Society of Professional Engineers.
- Baine, Celeste. 2007. *The Musical Engineer: A Music Enthusiast's Guide to Careers in Engineering and Technology.* Springfield, OR. Bonamy Publishing.
- Baine, Celeste, Cox, Cathi. 2007. *Teaching Engineering Made Easy: A Friendly Introduction to Engineering Activities for Middle School Teachers.* Springfield, OR. Bonamy Publishing.
- Catalyst. 2004. The Bottom Line: Connecting Corporate Performance and Gender Diversity. Available online at: http://www.catalyst.org/knowledge/titles/title.php?page=lead_finperf_04.
- CIA (Central Intelligence Agency). 2001. Long-Term Global Demographic Trends: Reshaping the Geopolitical Landscape. Available online at *https://www.cia.gov/library/reports/general-reports-1/Demo_Trends_For_Web.pdf.* Accessed September 28, 2008.

- College Board. 2004, 2006. Available online at: *http://www.collegeboard.com/ap/library/.*
- Cotton, K. 1996. School Size, School Climate, and Student Performance. School Improvement Research Series (SIRS), Close-up #20. Portland, Ore.: Northwest Regional Educational Laboratory. Available online at: *http://www.nwrel.org/scpd/sirs/10/c020.html.*
- Decker, P., D. Mayer, and S. Glazerman. 2004. The Effects of Teach For America on Students: Findings from a National Evaluation. Princeton, N.J.: Mathematica Policy Research, Inc. Available online at: *http://www.mathematica-mpr.com/publications/pdfs/teach.pdf.*
- EWE (Extraordinary Women Engineers). 2005. Extraordinary Women Engineers: Final Report. Available online at: *http://www.engineeringwomen.org/pdf/EWEPFinal.pdf.*
- Guess, Andy. 2008. *Enrollment Surge for Women.* Inside Higher Ed. Washington, DC: Available online at: *http://www.insidehighered.com/news/2007/08/07/enrollment.*
- Harris Interactive. 2004. American Perspectives on Engineers and Engineering. Conducted for the American Association of Engineering Societies. Final report February 13, 2004. Available online at: *http://www.aaes.org/harris_2004_files/frame.htm.*
- International Technology Education Association (ITEA). 2008. *Advancing Excellence in Technological Literacy.* Reston, VA.
- International Technology Education Association (ITEA). 2008. *National Standards for Technological Literacy.* Reston, VA.
- Koehler, C., Faraclas, E., Giblin, D., Kazerounian, K., and Moss, D., 2006. "A State by State Analysis of Engineering Content in State Science Frameworks: Toward the Goal of Universal Technical Literacy." *Journal of Engineering Education.*
- Lewin, Tamar. July 25, 2008. Math Scores Show No Gap for Girls, Study Finds. The New York Times. Available online at: *http://www.nytimes.com/2008/07/25/education/25math.html?_r=1&scp=1&sq=%2b%22National+Science+Foundation%22&st=nyt&oref=slogin*
- Margolis, J., and A. Fisher. 2002. Unlocking the Clubhouse: Women in Computing. Cambridge, Mass.: MIT Press.
- Meade, S.D. and W.E. Dugger. 2004. "Reporting on the Status of Technology Education in the U.S." The Technology Teacher 64(2): 29-35

- NAE (National Academy of Engineering). 2005. *Engineering Research and America's Future: Meeting the Challenge of a Global Economy*. Washington, DC: National Academies Press.
- NAE (National Academy of Engineering). 2005. *Educating the Engineer of 2020: Adapting Engineering Education to the New Century*. Washington, D.C.: National Academies Press.
- NAE (National Academy of Engineering). 2008. *Changing The Conversation: Messages for Improving Public Understanding of Engineering*. Washington, D.C.: National Academies Press.
- NASA Workforce Report. 2007. Available online at http://nasapeople.nasa.gov/workforce/.
- NCES (National Center for Education Statistics). 2001. The Nation's Report Card: Mathematics 2000. NCES 2001–517, by J.S. Braswell, A.D. Lutkus, W.S. Grigg, S.L. Santapau, B.S.-H. Tay-Lim, and M.S. Johnson. Washington, D.C.: Office of Educational Research and Improvement, U.S. Department of Education.
- National Council of Teachers of Mathematics (NCTM). 2000. Principles and Standards for School Mathematics. Reston, VA: NCTM.
- National Research Council and the National Academy of Sciences. 1996. National Science Education Standards. Washington, DC: National Academies Press.
- NSB (National Science Board). 2004. Science and Engineering Indicators 2004. NSB 04-01. Arlington, Va.: National Science Foundation. Available online at: *http://www.nsf.gov/statistics/seind04*.
- NSB. 2006. Science and Engineering Indicators 2006. Vol. 1, NSB 06-01; Vol. 2, NSB 06-01A. Arlington, Va.: National Science Foundation. Available online at: *http://www.nsf.gov/statistics/seind06*.
- NSF (National Science Foundation). 2003. New Formulas for America's Workforce: Girls in Science and Engineering. Available online at: http://*www.nsf.gov/pubs/2003/nsf03207/nsf03207_1.pdf*
- Singh, Manpal. 2005. Modern Teaching of Mathematics. Anmol Publications Pvt Ltd Daryaganj, New Delhi
- Sullivan, J.F., M.N. Cyr, M.A. Mooney, R.F. Reitsma, N.C. Shaw, M.S. Zarske, and P.A. Klenk. 2005. The TeachEngineering Digital Library: Engineering Comes Alive for K–12 Youth. In Proceedings of the ASEE Annual Conference, Session 3510, June 2005, Portland, Oregon. Available online at: www.teachengineering.com.

- USCB (U.S. Census Bureau). 2000. U.S. Census Bureau National Population Projections. Available online at *http://www.census.gov/population/www/projections/natproj.html.*
- Winstein, Keith J. July 25, 2008. Boys' Math Scores Hit Highs and Lows. The Wall Street Journal. Available online at: *http://online.wsj.com/article/SB121691806472381521.html.*

Programs, Initiatives & Classroom Curricula

AP (Advanced Placement Programs)
http://www.collegeboard.com/student/testing/ap/about.html

Amatrol: integrated technical learning systems
http://www.amatrol.com

Autodesk Design Academy
http://usa.autodesk.com/adsk/servlet/index?siteID=123112&id=3268568

CEEO (Center for Engineering Educational Outreach), Tufts University
http://www.ceeo.tufts.edu

Infinity Project
http://www.infinity-project.org

Intel Innovation in Education
http://www97.intel.com/education/

IB (International Baccalaureate)
http://www.ibo.org/ibo

JASON Foundation for Education
http://www.jasonproject.org/home.htm

PLTW (Project Lead the Way)
http://www.pltw.org/

SEEK-16 (Strategies for Engineering Education K-16 Summit)
http://www.howardcc.edu/seek-16

Design and Discovery
http://www97.intel.com/discover/DesignDiscovery/DD_Research/

EXITE (EXploring Interests in Technology and Engineering, IBM)
http://www.ibm.com/ibm/ibmgives/grant/education/camp.shtml

Intel NWSE (Northwest Science Expo)/ISEF (International Science & Engineering Fair)
http://www.nwse.org/

MESA (Mathematics, Engineering, Science Achievement)
http://mesa.ucop.edu/home.html

Oregon Building Congress Math and Science Workshops
http://www.obcweb.com

ORTOP (Oregon Robotics Tournament Outreach Program), OUS
http://www.ortop.org

Saturday Academy - PSU / OHSU
http://www.saturdayacademy.org

SMILE (Science and Mathematics Investigative Learning Experiences), OSU
http://smile.oregonstate.edu

STARBASE (Science and Technology Academies Reinforcing Basic Aviation and Space Exploration)
http://starbasedod.org

Techno Science Supersite
http://www.technosciencesupersite.org/

Youth Exploring Science - YES
http://www.youthexploringscience.org

Programs & Initiatives, Co-curricular, National

American Society of Mechanical Engineers Diversity Outreach
http://www.asme.org/communities/diversities/bdo/

Discover Engineering
http://www.discoverengineering.org/

Dream It. Do It. Campaign
http://www.dreamit-doit.com/

FIRST (For Inspiration and Recognition of Science and Technology)
http://www.usfirst.org/

Girl Power 21st Century
http://www.celt.sunysb.edu/gp21

Girls Go Tech
http://www.girlsgotech.org/

Girls Research Our World
http://www.ksu.edu/grow

Graduates Linking with Undergraduates in Engineering (GLUE) at UT Austin
http://www.engr.utexas.edu/wep/glue/

Lemelson-MIT InvenTeams
http://web.mit.edu/inventeams/

MATHCOUNTS
http://www.mathcounts.org/

National Engineers Week
http://www.eweek.org/

PNNL (Pacific Northwest National Lab) Science and Engineering Education
http://science-ed.pnl.gov/

TexPREP & SAPREP (The Texas Pre-freshman Engineering Program)
http://www.prep-usa.org/portal/saprep

Women in Engineering Proactive Network
http://www.wepan.org/

Organizations

ASEE (American Society for Engineering Education)
http://www.asee.org/

Advanced Technology Education Centers
http://www.atecenters.org/

BEC (Business Education Compact)
http://becpdx.org

CIESE (Center for Innovation in Engineering and Science Education)
http://k12science.ati.stevens-tech.edu

EESC (Engineering Education Service Center)
http://www.engineeringedu.com

EPICS (Engineering Projects in Community Service)
http://epicsnational.ecn.purdue.edu/

Future Scientists and Engineers of America
http://www.fsea.org/

IEEE (Institute of Electrical and Electronics Engineers)
http://www.ieee.org/portal/sitel

ITEA (International Technology Education Association)
http://www.iteaconnect.org

JETS (Junior Engineering Technical Society)
http://www.jets.org

NACME (National Action Council for Minorities in Engineering)
http://www.nacme.org/

NSBE (National Society of Black Engineers)
http://www.nsbe.org

NSTA (National Science Teachers Association)
http://www.nsta.org/

NWREL (Northwest Regional Education Laboratory)
http://www.nwrel.org/

OACTE (Oregon Association of Career and Technical Educators)
http://www.oregonacte.org/

OMSI - Oregon Museum of Science and Industry
http://www.omsi.edu/

PAVTEC
http://www.pcc.edu/pavtec/

RISE (Resources for Involving Scientists in Education)
http://www.nationalacademies.org/rise/

SECME, Inc
http://www.promorphus.com/secme/index.ph

SEMI High Tech U
http://www.semi.org/foundation

SHPE (Society of Hispanic Professional Engineers)
http://www.shpe.org

TEO (Technology Educators of Oregon)
http://www.teoregon.com

Women @ SCS (the School of Computer Science), Carnegie Mellon University
http://women.cs.cmu.edu/

Publications

ACM (Association of Computing Machinery) K-12 Taskforce
http://csta.acm.org/Curriculum/sub/ACMK12CSModel.html

ADVANCE: Increasing the Participation and Advancement of Women in Academic Science and Engineering Careers (NSF Grant Opportunity)
http://www.fedgrants.gov/Applicants/NSF/OIRM/HQ/05-584/Grant.html

Benchmarks for Science Literacy (AAAS publication)
http://www.project2061.org/publications/bsl

EngineeringK12 Center
http://www.engineeringk12.org/

"IBM Encourages Employees to Become Teachers" Washington Post Article
http://www.washingtonpost.com/wp-dyn/content/article/2005/09/16/AR2005091600674.html

Latino Students and the Educational Pipeline
http://www.educationalpolicy.org/pdf/LatinoIII.pdf

Losing the Competitive Edge (a report by the AeA)
http://www.aeanet.org/Publications/idjj_CompetitivenessMain0205.asp

MOS ERC (Museum of Science's Education Resource Center)
http://www.mos.org/doc/1369

Science for All Americans (AAAS publication)
http://www.project2061.org/publications/sfaa

TeachEngineering Digital Library
http://www.teachengineering.com/

Index

D

E

I

J

K

L

M

R

S

T

Other Engineering Publications by Celeste Baine

About the Author

Celeste Baine is a biomedical engineer, motivational speaker, the director of the Engineering Education Service Center and the award-winning author of 20 engineering publications. She won the 2005 Norm Augustine Award for Engineering Communications which is given to an engineer who has demonstrated the capacity for communicating the excitement and wonder of engineering; the 2004 American Society for Engineering Education's Engineering Dean Council's Award for the Promotion of Engineering Education and Careers; and, she is also listed on the National Engineers Week Web site as one of 50 engineers you should meet!

The Engineers Make a Difference Campaign Across America!

Learn how to make a difference.

Change the current landscape of engineering education.

Get involved!

www.engineersmakeadifference.com